Theme Book on Business Analytics

Table of Contents

www.misqe.org

Please note that page numbers in this theme book do not correspond to the page numbers of the articles as originally printed in standard issues of MIS Quarterly Executive. The articles are paginated in this book in chronological order of the original publishing date.

A research journal devoted to improving practice

The mission of MIS Quarterly Executive is to encourage practice-based research in information systems and to disseminate the results of that research in a manner that makes its relevance and utility readily apparent.

Editor-In-Chief
Dorothy E. Leidner
Baylor University

Past Editors-in-Chief

Jack Rockart, MIT Sloan School of Management, (2001-2005)   Carol Brown, Stevens Institute of Technology, (2009-2013)
Jeanne Ross, MIT Sloan School of Management, (2005-2008)

Editorial Offices

Publisher	Lead Architect	Production Editor	Managing Editor
Alan Dennis	**Ramesh Venkataraman**	**David Seabrook**	**Jordan B. Barlow**
Indiana University	Indiana University		Indiana University

MIS Quarterly Executive (MISQE) is a peer-reviewed research journal, affiliated with MIS Quarterly, that is devoted to improving information systems practice. MISQE is distributed on the web at www.misqe.org, and paper copies of the journal are available at an additional fee. For subscription information, please visit the MISQE website.

MISQE invites authors to submit papers based on in-depth research that provide rich stories, unique insights, and useful conceptual frameworks for information systems practice. Our target audience includes both practitioners and researchers – so that MISQE can stimulate ongoing discussions at the intersection of research and practice – but our primary focus is research that is immediately relevant and useful for practice. For additional submission information for authors, please visit www.misqe.org.

Foreword[1] ████████████████████████████

In the digital economy, data is one of an organization's most important assets; however, in recent years, the "look" of the data asset has evolved to become voluminous, multi-structured, real-time, ubiquitous, and interconnected. This contemporary appearance can be attributed to the influx of sensors, mobile devices, social medial, video streams, RFID, voice over IP, geocoding, telematics – and the advancement of supporting infrastructural and processing technologies.

Although the data asset's changing look is important for leaders to monitor and understand, it is even more important for leaders to comprehend data's changing role in organizations. Data is no longer an asset that we need to merely store, manage, and protect. Instead, data is an asset to exploit in ways that increase end consumers' willingness to pay for core goods and services; enable new business models; enrich an organization's understanding of players across its supply chain; improve strategic, tactical and operational decisions; and drive firm innovation.

Exploiting today's data is not straightforward. In this special theme book published by *MIS Quarterly Executive*, a diverse set of big data research articles offer insights into the changing look of data's managerial opportunities and challenges.

It is our hope that the findings and recommendations from these papers will be useful to managers who are facing challenges from big data while also hoping to seize the opportunities inherent in it. As a group, these articles will help managers think more broadly about big data in context, design big data governance structures with an eye towards innovation, make sense out of big data, and extract additional value from their big data.

Cynthia Beath (cbeath@mail.utexas.edu)
Jeanne Ross (jross@mit.edu)
Barbara Wixom (bwixom@mit.edu)

---

1 Adapted from the introduction to *MISQE*'s special issue on business analytics (Beath, C., Ross, J., and Wixom, B. "Editor's Comments," *MIS Quarterly Executive* (12:4), 2013, pp. iii-iv). Used with permission.

VIGILANT INFORMATION SYSTEMS FOR MANAGING ENTERPRISES IN DYNAMIC SUPPLY CHAINS: REAL-TIME DASHBOARDS AT WESTERN DIGITAL[1]

Robert Houghton
Western Digital Corp

Omar A. El Sawy
University of Southern California

Paul Gray
Claremont Graduate University

Craig Donegan
Western Digital Corp

Ashish Joshi
Western Digital Corp

Executive Summary

This article describes how Western Digital (WD), a global hard-drive manufacturer that supplies over 100,000 hard drives a day, built a vigilant information system (VIS) that includes both sensing and responding capabilities. The system includes an underlying layer of business intelligence applications that analyze data from numerous sources, and management dashboards that automate the alerting process and provide the means for responding. Its operational costs have been reduced almost 50% due both to the VIS and to revamping WD's business processes so that the right people are alerted and have the means to respond correctly and quickly.

Seven lessons were learned from this effort:

1. *Design the real-time management dashboards to be the nerve center for managing the enterprise.*
2. *Plan and schedule the coordination among teams to use the dashboards to manage enterprise-wide.*
3. *Build a learning loop around each OODA (observe, orient, decide, act) loop to foster group learning because the faster the loop, the more important the learning reviews*
4. *Match the time latency of each OODA loop to the organization's needs and capabilities to become truly vigilant. Do not indiscriminately chase zero latency.*
5. *Provide the building blocks for the "sense-and-respond" real-time enterprise through a vigilant information system and real-time management dashboards.*
6. *Justify vigilant information systems on a basis other than return on investment.*
7. *Make implementation of an enterprise-wide VIS a management initiative (not a technology initiative) because it requires "active, collaborative engagement" from all top management to instill the needed organizational transformation.*

VIGILANT INFORMATION SYSTEMS [2]

In a cost-conscious and turbulent economy, operating effectively in a lean and high-velocity supply chain is demanding, especially for high-volume suppliers whose large customers change their requirements often. In such dynamic supply chains, *vigilant information systems* are needed to respond quickly.

To be vigilant means to be alertly watchful. A vigilant information system (VIS) includes both sensing and responding capabilities. *Sensing*—to detect changes and enhance managerial visibility from the factory shop floor to corporate headquarters—comes through real-time dashboards with automated alerting. *Responding* comes through capabilities that help decision makers at each organizational level reach decisions and take action.

The dashboards at Western Digital are called "real time," which means they are "sufficiently vigilant for the process being monitored." In other words, "real

[1] Cynthia Beath was the accepting Senior Editor for this article.
[2] We are indebted to the following people for their contributions to WD's vigilant information systems, for managing the dashboard project, and for the contributions to this paper: John Coyne, Senior Vice President, Worldwide Operations; Chee Pang Sin, Director of Information Technology, Western Digital; Asia; Tom McDorman, Vice President and Managing Director, WD Asia; and Richard Chang, Staff Data Base Administrator, Asia.

time" for the factory means "as close to real time as possible," while "real time" for executive management means "once the information has been validated and synchronized among data feeds so that noise has been filtered out."

To understand VIS, it's helpful to understand how these systems differ from traditional information systems, and how the concept of OODA loops is useful in designing VIS.

How Vigilant Systems Differ From Traditional Systems

Vigilant information systems integrate and distill information and business intelligence[3] from various sources to detect changes, initiate alerts, assist with diagnosing and analyzing problems, and support communication for quick action.[4]

A vigilant information system differs from a traditional system as shown in Figure 1. In a traditional information system, a user initiates a process, by, say, querying an application, causing a database to be accessed. As shown, the database—whether it stores transactions, events, or other data— is passive.

In contrast, in a vigilant information system, the system initiates the process. As shown, the database is active. Each time its data is updated, the data is reanalyzed. If preset conditions are met, the system alerts the user. Thus, vigilant information systems provide sense-and-respond capabilities.[5] VIS are meant to provide proactive mechanisms and are designed from strategic and operational plans.

OODA Loops and Their Applicability to Vigilant Information Systems

The requirements for vigilant information systems can be viewed as coming from the concept of an OODA loop (Observe, Orient, Decide, Act). See Figure 2.

US Air Force Colonel John Boyd developed the concept of the OODA loop in 1986.[6] He wanted to understand how fighter pilots flying aircraft with inferior maneuverability won air combat engagements (dog

fights) against pilots with superior aircraft. He found that the winning pilots compressed the cycle of activities in a dog fight and completed them more quickly than their adversaries. Boyd's OODA loop included:

- *Observe* (see the situation and adversary)
- *Orient* (size up the vulnerabilities and opportunities)
- *Decide* (choose the combat maneuver to take)
- *Act* (execute the maneuver)

In 1990 and 1993, respectively, Stalk and Hout[7] and Haeckel and Norton[8] converted the idea to business use. Haeckel and Norton's four activities are:

- *Observe* (see change signals)
- *Orient* (interpret the signals)
- *Decide* (formulate an appropriate response)
- *Act* (execute the selected response)

When changes occur in the business environment, enterprises that can complete OODA loops faster than competitors improve their ability to survive. OODA loops should not only be executed quickly but they should also be flexible and responsive to environmental changes.

Western Digital's corporate managers use OODA thinking to react quickly to customer changes. In fact,

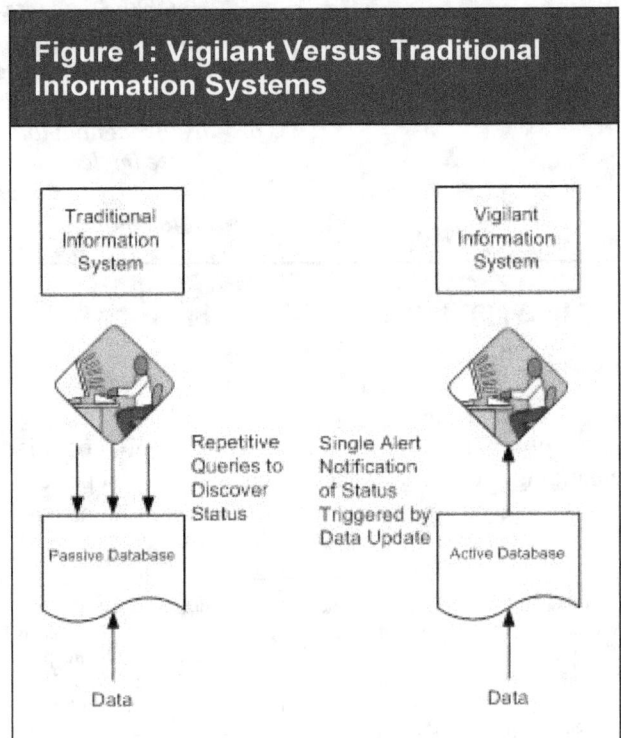

Figure 1: Vigilant Versus Traditional Information Systems

---

[3] Business intelligence is granular information about the business and its supply chain that line-of-business managers seek when they are analyzing key performance metrics of their enterprise.

[4] Walls, J., Widmeyer, G., and El Sawy, O. "Building an Information System Design Theory for Vigilant EIS," *Information Systems Research*, March 1992, pp. 36-59.

[5] Haeckel, S. *Adaptive Enterprise: Creating and Leading Sense-and-Respond Organizations*, Harvard Business School Press, 1999.

[6] Boyd, J. Patterns of Conflict, Unpublished manuscript, USAF, 1986. Also see Curts, R. and Campbell D. "Avoiding Information Overload through the Understanding of OODA Loops," *Proceedings of the Command & Control Technology Research Symposium*, 2001.

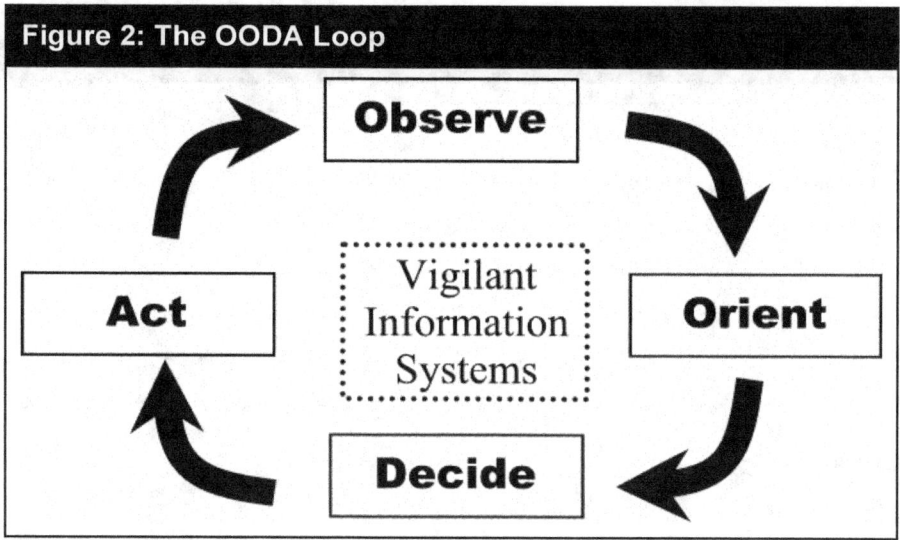

Figure 2: The OODA Loop

the goal is to initiate a change in the factories within the same work shift in which WD receives requests from customers; being the fastest to respond to new customer requirements in the hard drive business can increase WD's market share.

The OODA loop translates into the following four requirements for VIS:

- **Capabilities for observing:** Provide visibility into the critical business processes in the enterprise's supply chain. Capture key performance indicators (KPI's) in real time. Integrate information from various sources and systems;

- **Capabilities for orienting:** Provide graphical dashboards that display data. Send alerts to managers. Permit drilling down in data. Allow users to slice-and-dice the data. Provide traffic-light alerts. Report trends.

- **Capabilities for deciding:** Analytics for asking "what-if" questions. Descriptive statistics. Time series comparisons.

- **Capabilities for acting:** Architectures for communicating decisions quickly to pre-specified others to take action. Follow-up tracking.

Western Digital Corporation applied these concepts to build their vigilant information system.

WESTERN DIGITAL AND ITS BUSINESS CHALLENGES

Western Digital is a $3 billion global designer and manufacturer of high-performance hard drives for desktop personal computers, corporate networks, enterprise storage, and home entertainment applications. Founded in 1970, WD sells its hard drives to system manufacturers, resellers, and retailers. Headquarters are in Lake Forest, California, about 50 miles south of Los Angeles. Its manufacturing facilities are in Malaysia and Thailand, and it has distribution centers in Europe. WD employs about 18,000 people worldwide.

WD's top five business challenges are:

1) Constantly changing customer requirements for more storage space, faster access, and better performance

2) A fiercely competitive global industry that exerts pricing pressures

3) Avoiding business disruption, product returns, excess inventory, and bad scheduling

4) Short product lifecycles and rapid obsolescence

5) The need for extremely high quality and reliability in its products

In the early 1990s, the hard drive industry consisted of over 11 manufacturers. It now involves 3 to 5, depending on the product line. WD has not only survived but excelled, becoming the third largest volume producer. In 2002, unit volume rose 30% over 2001 to 29 million drives, and gross margins improved—even though 2002 was one of the toughest years in the IT industry.

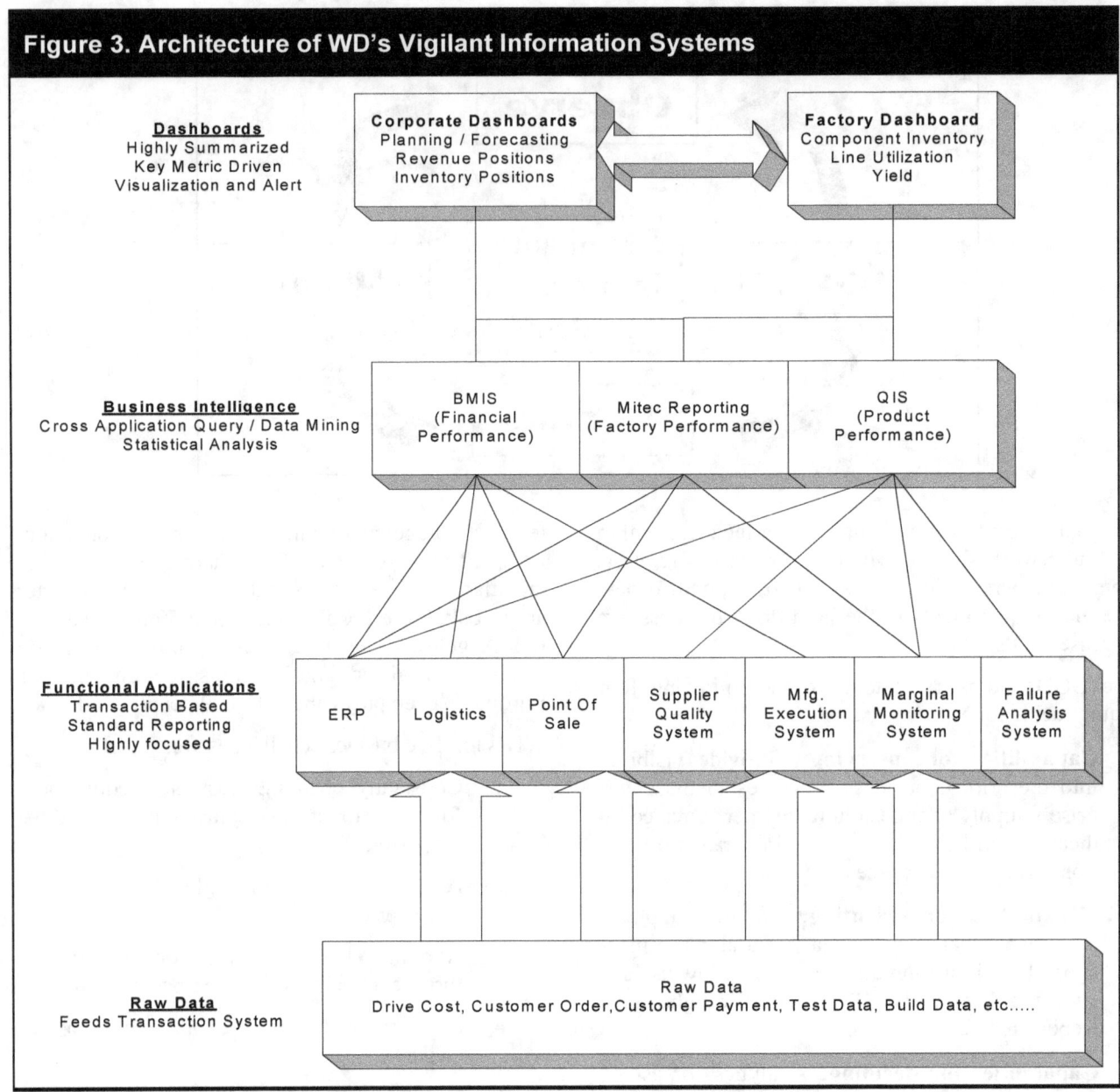

Figure 3. Architecture of WD's Vigilant Information Systems

As part of WD's survival strategy, management demanded a new mode of information delivery. First, they wanted the ability to react more quickly to changes. Second, they wanted integrated information so that they could manage enterprise-wide in a "follow the sun" manner.[9]

Like many enterprises, information at WD used to be difficult to consolidate because there were no single sources of data. When users ran ERP reports using different filters, they received different results. The data was not accurate nor was it current. Management and end users had no easy way to see trend data, understand the current state of the business, or use a system to take quick action.

Delivering the needed capabilities meant creating an IT architecture that would support not only 24 x 7 system availability but also integrate applications and data. Managers and analysts needed real-time visibility of data in easy-to-use formats. They needed to be notified of changes, when they occurred, in key performance indicators, delays, or supply-demand imbalances. And they needed enough depth of information that different organizational levels could use the same tools. Finally, they needed a way to take action quickly because that is when action makes the greatest difference.

---

[9] "Follow the sun" refers to passing work or information across multiple time zones as the workday closes in one and opens in another.

The solution to these needs is WD's vigilant information system and its real-time management dashboards.

WESTERN DIGITAL'S VIGILANT INFORMATION SYSTEM

The VIS is complex, so its description is divided into four sections: its overall architecture, three foundation capabilities, revamping Western Digital's business processes, and the management dashboards.

The VIS Architecture

Figure 3 shows a four-layer schematic view of the architecture of WD's VIS. Starting from the bottom, in Layer 1 is the raw data, which comes from various sources. That data flows into numerous functional applications (ERP, logistics, and so on) in Layer 2 (observe). Business intelligence systems in Layer 3 analyze each new piece of data to determine whether or not it is within its preset boundaries (orient). Out-of-bounds data initiates an alert to Layer 4, the dashboards, at the top (decide-and-act).

Three Foundation Capabilities

Three IT capabilities form the foundation for WD's VIS: the ERP system, the data warehouse, and the Quality Information System (QIS). These systems are considered foundations because together they capture and integrate the data needed for the VIS.

ERP was implemented in 1997, giving managers data about enterprise operations they had never had before. They could, for example, tie together requisitions, purchase orders, sales orders, production runs, and invoices. And they had on-line access to day-to-day workflow and manufacturing processes.

The data warehouse was implemented in 1999 and integrated product data from 12 disparate legacy systems.[10]

The Quality Information System (QIS), also implemented in 1999, is in the business intelligence layer and integrates data to provide insights into quality. It is used to trace each hard drive manufactured at WD through its entire lifecycle, including components, manufacturing, testing, shipment, and returns. WD's

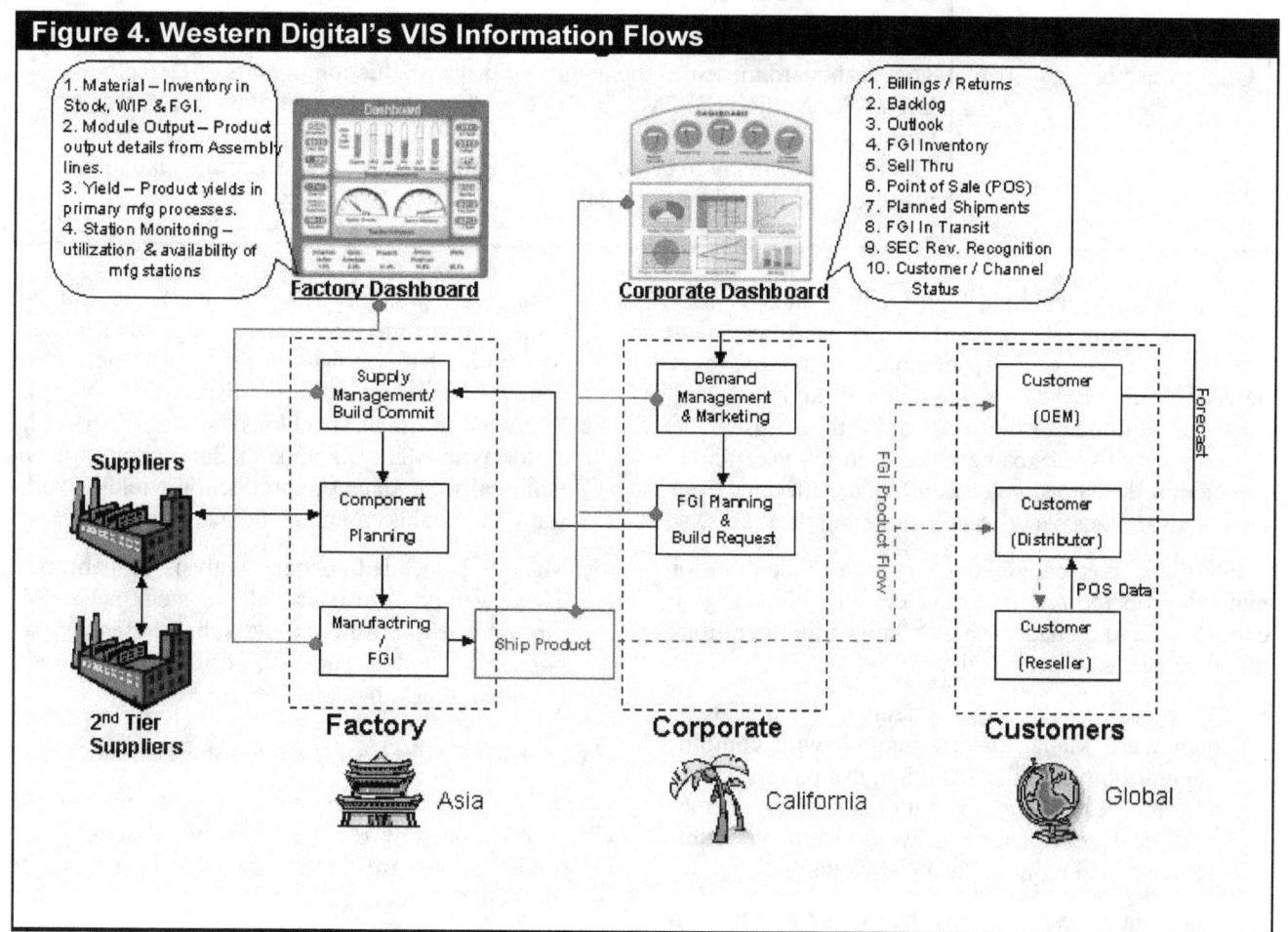

Figure 4. Western Digital's VIS Information Flows

Figure 5. Descriptions of the Factory Dashboards

Yield dashboard	Yield is the percentage of units that pass inspection. This dashboard shows yields by product, capacity, cache family, and station level. It also reports cycle time of key processes and at a key manufacturing station, Servo Track Writer (STW). (See Figure 6.)
Material (inventory) dashboard	This dashboard tracks factory inventory, including the receiving warehouse, work in progress, engineering locations, and finished goods inventory. Targets are set for each location based on the daily manufacturing schedule.
Production output dashboard	This dashboard tracks assembly line output, including hard drive and printed circuit board assemblies. These outputs are compared with targets based on the manufacturing schedule. Because the assembly station is the beginning of the drive manufacturing process, any delay or slip in output can affect the entire manufacturing and shipping operation.
SPT monitor dashboard	This dashboard monitors activities and availabilities of one of the longest processing periods and most important stations in the drive manufacturing process, Single Plug Tester (SPT). Individual drives are tested in SPTs for hours, and hundreds of SPTs are on the factory floor, loaded with drives. SPTs can become bottlenecks unless used to their maximum capacity. This dashboard gives real-time visibility of each SPT, including utilization and the products being run.
Quality dashboard	This dashboard measures the quality of drive production in units of Defective Drives Per Million (DPPM). DPPM targets are based on products and customers. Problems with components or tools on the manufacturing line introduce defects. The longer the delay in detecting a problem, the more extensive (and expensive) the effort to repair the affected drives.

objectives are to maintain high quality, uncover root causes of product failures, and improve future versions of products. QIS can pinpoint component-level defects before hard drives are shipped, and can trace back from failures in the field to the root cause. The data captured by QIS can also be used to make future production decisions, when combined with data from the data warehouse and R&D's modeling databases.

While these three capabilities form the foundation for managing operational performance, they did not give executives and managers the visibility into operations that they needed. Specifically:

- A number of legacy systems remained non-integrated. Management struggled with combining real-time and historical data from 30+ systems to make strategic and operational decisions. Queries and report generation were expensive, cumbersome, and required many consultants.

- The data refresh rate was inadequate and uneven across systems. Outdated data led to different re-

sults (there was no "single truth") and did not support real-time decision making. Management was without the real-time information needed in demand management and distribution (that is, forecasts, billings, backlogs, sell-through). The factories in Malaysia and Thailand could not obtain real-time data on production yields, workstation availability, and component inventories.

- Managers needed better analysis capabilities. They wanted a user-friendly system accessible through the Web that would support standard reporting, detailed, ad-hoc, drill-down queries, graphical reporting, and executive summaries.

Revamping WD's Business Processes

WD's senior management team, which includes the CIO, looked at how real-time dashboards could best be used. They agreed that the dashboards alone wouldn't change decision making.

Take, for example, the 5-hour daily production meetings (8:30 a.m. to 1:30 p.m.) where factory management decided what needed to be produced. Not only did these meetings manage production inadequately but corporate management in California did not know of the decisions made in these meetings. Management realized these meetings could be much shorter (less than 2 hours, in fact) if the participants used the dashboard data. Furthermore, the executives in California could work with updated production information.

Top management drafted new business policies and processes to put the VIS to work. Three new policies were deemed critical:

- **Align time-based objectives across the enterprise:** WD had to translate strategic enterprise goals into measurable, time-based operational objectives for each department. The result would be consistent metrics.

- **Capture key performance indicators (KPIs) in real time:** To improve corporate performance, WD needed real-time monitoring—horizontally across organizational groups and vertically within business units. With real-time KPIs, teams could analyze them across groups and business units.

- **Foster cross-team collaborative decision making:** The dashboard environment would need to enable joint decision making and collaborative working across teams, departments, enterprises, and geographic areas. Achieving such collaboration took months because the geographically dispersed teams had to decide what information they needed to hand off to R&D, corporate planning, new production groups, the factories, distribution, and customers—so that they could be "virtually there" via their dashboards. This concept of being "virtually there" was new to some WD executives.

In WD's VIS, the dashboards become managers' eyes and ears into operations. These policies aimed to ensure that decisions and actions were coordinated.

The Management Dashboards

Two real-time dashboard information systems were developed: one for the factory and one for demand planning, distribution, and sales information (the corporate dashboard).

The factory dashboards were custom-developed in-house and rolled out in late 2000. They are used to monitor such quantities as yields, quality, and production output. The corporate dashboards were developed from off-the-shelf software (Business Intelligence Product Suite from Cognos) and rolled out in 2001. They are used to monitor billings, sell-through, weeks of inventory, and more.

The dashboards tap into WD's information flows as shown schematically in Figure 4.[11] The four types of factory dashboards and ten types of corporate dashboards are shown at the top. The VIS and its information flows are in the middle. The three constituencies (at the bottom) are the factories in Asia, corporate offices in California, and customers around the globe. Information flows among these components as follows, going from right to left.

- Forecasts of customer demand from WD distributors and original equipment manufacturers (OEMs), on the right, flow into the central corporate business management information system (in the middle).

- Corporate combines these demand forecasts with feedback on production and inventory levels to create build requirements: how many units of which type are to be produced when.

- Based on the build requirements, each factory (on the left) determines its supply requirements and commits to building the hard drives. This production data is used for component planning by both first-tier and second-tier suppliers (on the far left). The product is then manufactured and ready for distribution

- Products shipped to customers provide data feedback both to the factory and to corporate.

The Factory Dashboards

The manufacturing and engineering staff who run a factory face tight requirements. When working near capacity, a factory can produce as many as 100,000 hard drives a day. Achieving this level of production requires that the production line not be shut down and that disks not require rework. Both cause significant economic loss. Furthermore, to remain competitive, WD continually improves its manufacturing process. An improved process must be monitored to make sure the modifications actually make things better.

---

[11] The dashboards shown at the top of Figure 4 are conceptual, not actual.

Figure 6: Yield Dashboard

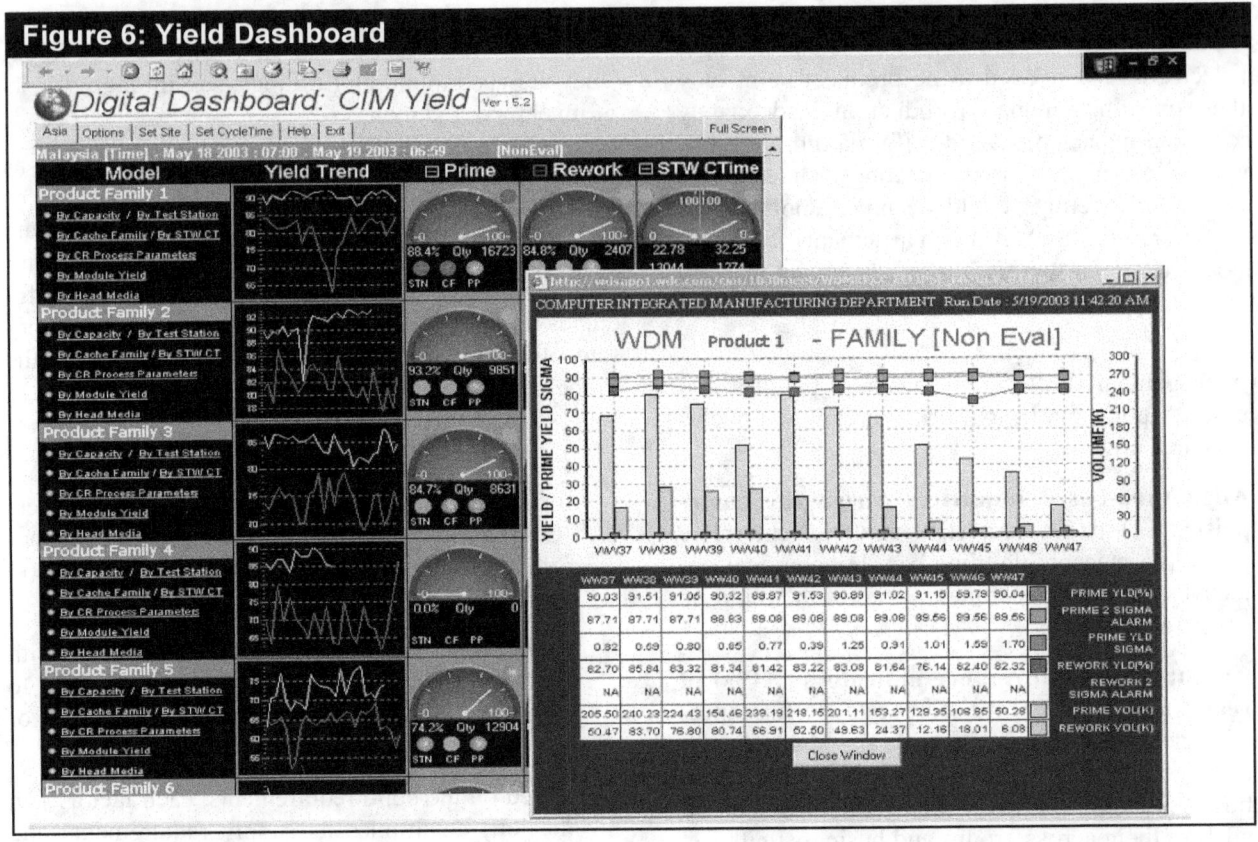

To assist factory staff in meeting these demands, the IT department built five factory dashboards. The four core requirements for the factory dashboards were to:

1) *Show KPIs*: Show the health of the factory by providing near-real-time, graphical views of KPIs.

2) *Display metrics*: Show when a KPI goes below 2 sigma of its allowable value.

3) *Allow drill down*: Give staff ways to drill down on each KPI to find the source of a problem.

4) *Issue alerts*: Automatically issue alerts to the individuals responsible for a KPI so they can initiate damage control.

The final requirement, for automated alerts, distinguishes these dashboards from typical executive information systems. Unanticipated events can cascade quickly, requiring fast response. Hence the need for automated alerts.

The five dashboards are yield, material, production output, station monitoring, and quality.[12] Each is de-

scribed in Figure 5, and the yield dashboard is shown in Figure 6.

Each parameter on a dashboard is assigned a "target value," which, when hit, causes an alert to be issued, often to the pager of the manager in charge of the station or product. The manager then uses a specific Decision Support System (DSS) to analyze the problem. As the vice president and managing director for WD Asia noted:

> *Our factory dashboards provide us with a 'virtual control room,' which, at a few moments' glance, tells us where our trouble spots are so we can respond much more rapidly to issues than before, and we can do it from practically any place in the world.*

The Corporate Dashboards

Ten corporate dashboards were created:

1) Billings and returns

2) Backlog

3) Outlook

4) Finished goods inventory

5) Distributor inventory and sell-through

---

[12] The Quality dashboard is not shown in Figure 3.

Figure 7. Distributor and Sell Through Report

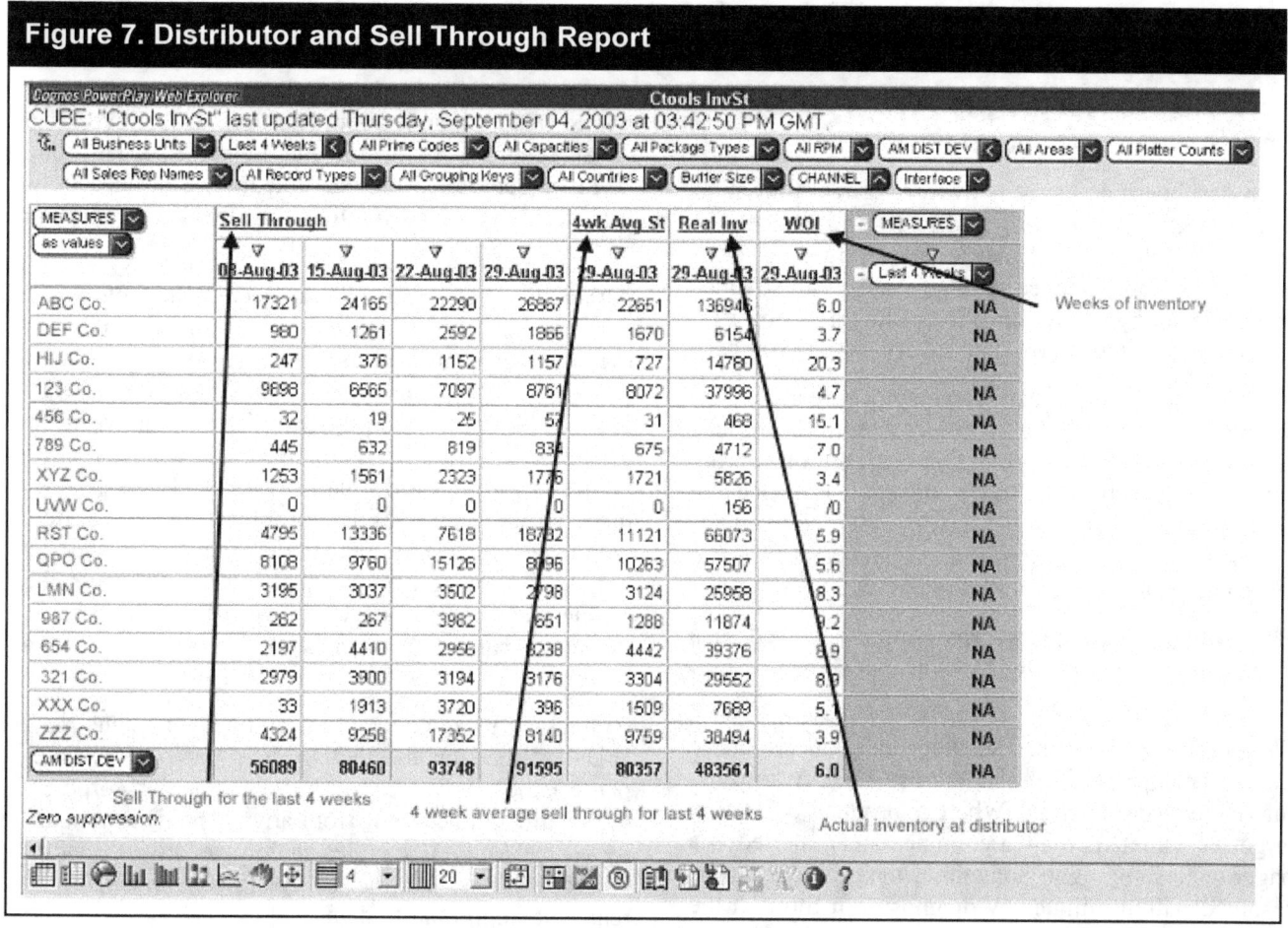

6) Point of sale

7) Planned shipments

8) Finished goods in transit

9) Revenue recognition

10) Customer/channel status

The dashboards are designed to accommodate all levels of management by varying the level of information aggregation. Each dashboard has its own "data cube," which houses the data for its displays. These data cubes are subsets of WD databases.

An example of one dashboard is the Distributor Inventory and Sell-Through Dashboard. "Sell-through" refers to goods sold through distributors. Figure 7 shows this dashboard. Before this dashboard was available, it was difficult and time consuming to find out how much finished goods inventory distributors had, how many finished goods were sold through the previous week, and how many weeks of inventory they had. The dashboard displays all this information on one screen.

The dashboard required changing data collection. Formerly, distributors reported sell-through and inventory at the end of each week. That data was entered into regional databases and then consolidated at corporate, where it was imported into an Excel spreadsheet. About one hour later, paper reports were created and distributed, unless errors and new categories caused delays.

Now, inventory and sell-through data is collected automatically into one repository. The data cube, which is used to tabulate the KPIs and present the data, is refreshed every 20 minutes during primary collection times, and once an hour for the remainder of the week. These frequent refreshes help detect new trends. New categories are automatically included, quarterly comparisons are also automatic, and a single e-mail notifies employees of the availability of the report.

Implementation of the dashboards encountered all the usual problems encountered in decision support and executive support systems: poor data, inadequate security safeguards, and user uncertainties about what they really wanted.

How the VIS Accelerates WD's OODA Loops

Each dashboard contains its own set of real-time metrics and KPIs for tracking and analyzing critical operations. Each KPI and metric has a target performance level and a variance setting (some set in advance and some set by the system). Exceeding a setting triggers an alert to the appropriate supervisor or manager.

A well-designed dashboard can help people accelerate the OODA loops of the processes monitored. It can also accelerate OODA loops that span multiple processes and departments. So, to manage the fast cycle environment of its supply chain, WD coupled its dashboards with both performance management and learning, using a nested structure of OODA loops.

The factory and corporate dashboards are used in a three-level nesting of OODA loops, as shown in Figure 8.

Shop floor OODA loop. The shop floor supervisors in the factory operate in the innermost OODA loop using factory dashboards. When a product or station target is violated, they are immediately alerted via pager or flashing light. Sometimes the problem can be resolved within minutes by diagnosing it through the dashboard.

Factory OODA loop. Production managers operate at the next higher level because they need a broader view of the factory—such as seeing multiple product lines. They also receive alerts when their targets are out of range. But a more important aspect of their job is using the factory dashboards, with a different set of KPIs, to perform "health checks" on the operational performance of the factory (that is, determining that things are working as they should). The health check is analogous to a medical health check, which measures vital signs that indicates whether or not critical body functions are within normal limits.

The production managers use the factory dashboards in more of a learning mode. In their daily production meeting, they use the factory dashboards to analyze the previous day's performance and discuss ways to improve processes or avoid problems encountered the day before.

Because of the real-time nature of the data, problems already handled by the factory's shop floor supervisors are filtered out, minimizing the information overload on the production managers. The managers can

more easily see the unresolved critical problems, which quickens their OODA loop.

Corporate OODA loop. Like the factory loop, the corporate OODA loop is about learning. It also involves a "health check"—but of the entire enterprise. This loop involves senior executives, and the resulting actions typically have broad implications, for the corporation and the factories.

Although not electronically connected, the factory and corporate dashboard systems are connected through the data they share and the communications and interactions of the managers who use them. The dashboards reduce the physical distances between the factories in Thailand and Malaysia and corporate management in California. As a result, people can meet virtually and resolve problems quickly. Often, employees or teams send references to a specific screen to another team so that both can discuss the problem and make a decision. To make this possible, managers can give permission for others to see a screen.

By making these connections and conversations easy, WD has sped up the OODA loops across the company. People who need either a factory or corporate dashboard can access it from anywhere. This capability very effectively couples senior executive decisions with operations, and does so by drawing on consistent, real-time, high-quality data.

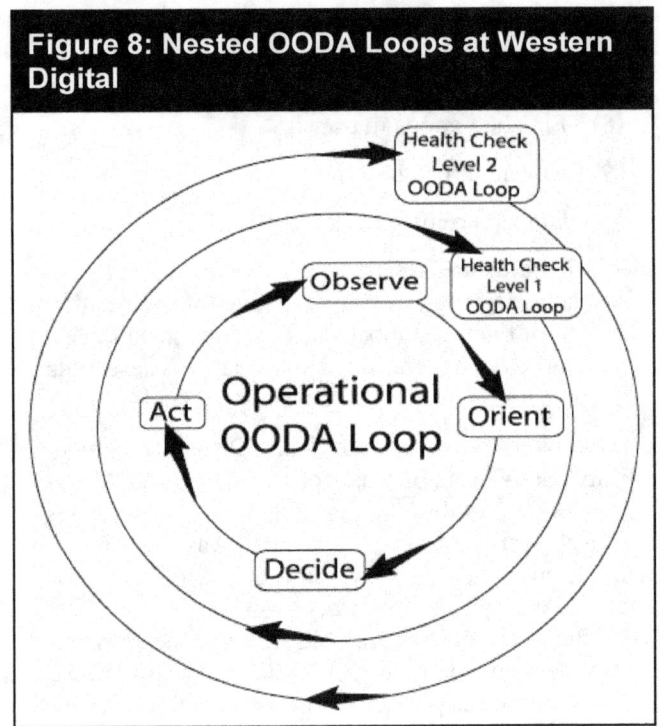

Figure 8: Nested OODA Loops at Western Digital

Dashboard use has not lead to micromanagement by top management, though. In fact, micromanagement was more prevalent before the dashboards. Previously, the shop floor supervisors, production managers and supply planners defined their own data in their "silo." The data came from redundant sources and often did not agree, if requested at different times. The dashboards give everyone the same data. Micro-intervention isn't needed because the data is more reliable.

THE BUSINESS IMPACTS OF WESTERN DIGITAL'S VIS

More than 225 managers and professionals at all levels of the company now use the dashboards. Their use changed how the OODA loops are managed, resulting in both measurable gains and continuing savings. On a pure cost-savings basis, these systems have paid for themselves many times over. However, the real payoff has come from quickening the OODA loops and decision cycles in ways that have changed WD's strategic capabilities.

Cost Savings

Cost savings have come in three forms: better visibility, more efficient querying, and less information overload and faster decision making.

Better visibility. Over the last several quarters, the increased visibility of finished goods inventory has allowed WD to increase inventory turns from 22 to 29, decreasing inventory by $25M in 2002, for annual savings of almost $3M in inventory carrying costs. Those savings alone paid for the entire dashboard IT infrastructure (the dashboards and the related reporting (business intelligence layer)). When this solution was proposed in FY2000 under the name "Business Management Information Systems," estimated cost of capital and expense was approximately $1.2M—with an ROI of one year. By saving almost $3M in inventory carrying costs in 2002, that ROI was realized in less than one year.

Western Digital's margins have more than doubled over the three-plus years since the dashboards were introduced. Management attributes these margin gains in large part to improvements in supply chain management, data visibility, and demand planning made possible with the dashboards. In improving demand management, WD has, on occasion, even been able to recapture missing revenue.

More efficient querying. Before the dashboards, it cost $1500, on average, for highly paid database administrators to create one of the hundreds of cross-application or cross-database reports requested each year. Now, users create multiple views "on the fly." Report requests have dropped from 200 to 50 per quarter, saving about $900K a year. Furthermore, the dashboards have eliminated about 50% of the printed paper volume, with an estimated savings of $800K a year. Custom reports now take an average of 10 minutes for managers to create, and they can easily combine real-time and historical data.

Less information overload and faster decision making. Executives, managers, and professionals now detect and analyze critical operational problems differently. Production supervisors and engineers no longer need to crunch data from several systems to measure KPIs. The dashboards feed this information to them. They can focus on performing the detailed analysis of trigger violations. Much of the noise and "unreal" problems are filtered out by the system. As a result, daily production meetings at the factories now take 1.5 hours, on average, rather than 5 hours. These meetings involve some 15 supervisors and managers, so time savings alone translates to $350K a year—time better spent on other work

Strategic Advantages

Strategic advantages come from faster analysis and decision making, immediately available information, quicker reflexes, and faster OODA loops, enterprise-wide.

Faster analysis and decision making. Executives using the corporate dashboards to identify KPI problems experience much less information overload. Formerly, they received information in different formats at different times from different managers, making quick, robust analysis difficult. Now, they can make strategic decisions quickly because they no longer need to spend hours trying to interpret the data correctly.

The better business intelligence capabilities has helped focus management attention. For example, one WD executive described how one product line can be used to benchmark others:

> With the dashboard we get one up-to-the-minute snapshot showing a comparison of yields for 16 product lines in the last month on one screen shot or graph, rather than 16 spreadsheets that are difficult to compare. The difficulty is in showing relationships. Dashboards easily show us where the relationships intersect and help us to diagnose problems. Dashboards help provide knowledge rather than just information.

Immediately available information. With the dashboards, everyone sees the same information, anytime, anywhere and updated at the appropriate time intervals. Before the dashboards, a salesperson on the east coast visiting a customer at 8 a.m. could not obtain the most current information because the account manager on the west coast was not at work yet. As expressed by the Senior Director of Sales for Latin America,

> *The comfort zone that management can get to the data almost real-time, anywhere in the world, is practically priceless. We have had events where the dashboards were accessed from a customer site, and we were able to quickly explain an issue by confirming product shipments, quarter-to-date billings, and sell-through volumes on the spot. This type of service and information availability improved our quality image and market share.*

Quicker reflexes. The reaction time between receiving data and acting on it has shortened from hours, sometimes days, to minutes. Timely alerts to manufacturing supervisors for "out of control" situations has reduced waste on the production lines. Analysts can catch demand-supply imbalance problems as they occur (rather than the day after), and react quickly to resolve them and limit damage. As one account manager put it,

> *"Instead of sending cumbersome spreadsheets back and forth, we can look at the same data at the same time and make real-time decisions even when we are at different locations. We also eliminated the need to create the spreadsheets."*

Faster OODA loops enterprise-wide. While many of the business impacts discussed above resulted in measurable gains and millions of dollars in savings, the longer-term benefits are harder to assess quantitatively. Yet, they exist and are clearly strategic.

One benefit is that WD is continuously becoming a more agile competitor, as the OODA loops accelerate at all levels of the enterprise. Tactical OODA loops provide feedback and learning to operational management, which, in turn, feed the executive decision-making OODA loops. WD now ties together executive decision making, supply chain movements, and internal operations into a virtuous circle that can function effectively as various situations unfold. As a result, as an enterprise, WD can learn faster and act with more vigilance, even as the supply chain continues to speed up.

WD's supply chain recognizes the company's contributions. In May 2003, for example, WD was one of eight suppliers that won a WorldWide Procurement Supplier Award from Dell Corporation.

Matt Massengill, CEO of WD, said it this way:

> *"The alerts in the management dashboards allow us to drive quicker to root cause. Planning meetings are now about the performance in production or sell-through rather than about whether the data on spreadsheet A is consistent with B, C, and so on. We now spend more time making decisions based on the data."*

LESSONS LEARNED

The real-time dashboards and their underlying infrastructure have changed how WD manages operations, from the boardroom to the shop floor in the factory. The bottom-line effects of improved visibility, real-time information, and quicker OODA loops occur at all levels of the enterprise. WD is a faster-learning and more vigilant enterprise, so it can operate more leanly. Its ability to link executive decision making with operations in real-time through dashboards has given it a strategic edge as a responsive supply chain partner. Western Digital's experiences provide seven lessons.

Lesson #1: Design The Real-Time Management Dashboards To Be The Nerve Center For Managing The Enterprise.

WD's experience shows that management dashboards can become the central point for managing an extended enterprise vigilantly. By providing the means for human-computer interaction, they are a critical component of a VIS architecture. These dashboards may be glitzy, or not. Some may display mainly tables of text. They should provide instant access to information (i.e., observe). But they also should combine event-based alerts (i.e., orient) with company history. And the information should be sharable, to give people ways to track and manage (i.e., decide-and-act) activities and processes across functions.

Lesson #2: Plan and Schedule the Coordination Among Teams to Use the Dashboards to Manage Enterprise-Wide.

Dashboards can show real-time metrics, workflow executions, alerts, and various slices from information cubes from other dashboards, providing a holistic picture that allows teams to collaborate more quickly and make decisions. Often these teams are in different functional areas, different geographic locations, and responsible for different business processes. One purpose of providing the information via dashboards should be to eliminate the separation of people in space and time, thereby allowing managers to be "virtually" in the room. Another purpose is to identify problems and opportunities that could otherwise be missed.

This coordination is being accomplished at Western Digital by using the corporate dashboards as the focal point of meetings among managers from sales and marketing, distribution, and the factory. These meetings are not ad hoc, though. They are scheduled daily, weekly, monthly for specific purposes—quality review, production planning. The meetings provide the human coordination and intervention needed to deal with enterprise-wide problems. The goal is to act quickly in a coordinated fashion.

To manage enterprise-wide, Western Digital makes sure that people with multiple perspectives are always involved in an OODA loop.

Lesson #3: Build A Learning Loop Around Each OODA Loop To Foster Group Learning Because The Faster The Loop, The More Important The Learning Reviews And The More Frequent They Need To Be.

As cycle times become shorter, errors and exceptions occur more quickly as well. To keep up, enterprises need information systems and management processes that record incidents and retain their history, so that groups can review and learn from them.

Western Digital embeds group learning into its OODA loops through its organized "health checks" and review meetings, both of which take place around the dashboards. WD learned that the shorter the cycle times, the more frequent and important the "health checks." Each OODA loop needs its own learning loop, designed specifically for it. Dashboards and VIS need to be designed and managed to support this use, and the learning loops need to be part of the management culture.

Lesson #4: Match the Time Latency of Each OODA Loop to the Organization's Needs and Capabilities to Become Truly Vigilant. Do Not Indiscriminately Chase Zero Latency.

"Just because information is available in real-time does not mean that an enterprise can act instantaneously on it."[13]

At first sight, it would appear that achieving zero latency in any portion of an OODA loop is the Holy Grail. But Western Digital has learned that "real time" in practice actually should mean "being sufficiently vigilant for the OODA loop you are in"—not zero latency. On the sensing side, the incoming information must be validated; otherwise, there is a lot of "noise." Also, the various information feeds need to be synchronized. Both add time to the sensing portion of the loop, but improve the enterprise's ability to be vigilant.

At WD, the OODA loops around the manufacturing process are much faster and closer to real time than the OODA loops around demand management in the supply chain, because validating the information about sales and demand takes longer. Furthermore, some parts of the organization cannot or do not need to act in real time.

The most effective OODA loops provide fresh information when the organization needs to respond. The computer-based sensing portions support the human-based responding portions. There is no point in getting fast sensing (the observe and orient portions of the OODA loop) if no actions can be taken. Likewise, there is not point in being able to respond quickly (the decide-and-act portions) if there is no fresh information. Investing in sensing capabilities that provide information more frequently than an OODA loop requires may not be the best use of IT funds or management attention. When in concert, the two provide the building blocks for a sense-and-respond enterprise.

Lesson #5: Provide the building blocks for the "sense-and-respond" real-time enterprise through a vigilant information system and real-time management dashboards

IT vendors are currently pursuing positioning strategies for the "sense-and-respond" real-time enterprise as their hallmark: IBM touts "e-business on demand," HP embraces the "adaptive enterprise," and TIBCO

---

[13] Sawhney, M. "Real-Time Reality Check," *CIO Magazine*, March 2003, pp. 37-38.

sells "real-time publish and subscribe" architectures. However, as we understand it from the public information available and corporate white papers and advertisements, they are positioned mostly as solutions to technology platform problems, rather than management problems.

Understanding the principles of real-time dashboards, VIS, and the structure of OODA loops provide a more informed and richer path to providing the building blocks for the "sense-and-respond" enterprise. The WD experience shows that alerting capabilities can trigger very different real-time performance management in an enterprise. Such IT capabilities help to shape the quickness of the OODA loops for effective "sense-and-respond" working in dynamic supply chains, as well as the management processes around them.

Lesson #6: Justify vigilant information systems on a basis other than return on investment.

The real payoff of VIS comes from accelerating OODA loops in ways that change the capabilities of the enterprise. Oftentimes, though, such grand initiatives are difficult to cost-justify. Financial returns are hard to quantify and risks are high. The Internet bubble increased skepticism about IT-based transformation initiatives.

Sometimes the strategic opportunities of VIS can be quantified—such as the value of increasing the enterprise's ability to respond more quickly to changing customer needs or the value of limiting the enterprise's exposure to suppliers that cannot meet their commitments. CFOs accept estimates that translate these risks into money.

At WD, the CIO justified the IT investment in yet another way. To sway the CFO, he illustrated the "impact of not doing it" by pointing out that not investing would hamper WD's ability to meet its supply chain performance commitments.

With the increasing need to build security and privacy capabilities into information systems, CFOs are likely to accept such justifications because these two capabilities also cannot be justified by ROI.

Lesson #7: Make implementation of an enterprise-wide VIS a management initiative (not a technology initiative) because it requires "active,

collaborative engagement" from all top management to instill the needed organizational transformation.

Taking advantage of real-time dashboards and VIS requires transforming an enterprise's management processes. Such a transformation requires more than the usual "getting top management support." It requires the entire top management team to be actively collaborative because they are the ones who must model the process of using the new information and knowledge differently to manage and steer the enterprise. Use must start at the top.

At WD, CEO Matt Massengill was one of the first to use the corporate dashboard. He demonstrated his use by taking fresh data from the dashboard to his top management and staff meetings. He used the data to emphasize a point, generate lively conversation, and review numerous KPIs. He set an example for how to use the new goldmine of data to manage differently—thereby sending the message that all top management needed to change their way of managing as well.

In fact, Massengill adopted the dashboard so completely that he asked that some views be ported to his PDA. The IT team extracted key Billings to Date, Sales by Region, and Orders by Unit—the extracts most important to Massengill and other executives when they are on the road. Generally, the indicators are "all is well." But they are excellent for alerting the executives when production is below expectations. Such information can make all the difference between "go ahead and board the plane" or "get to a phone."

THE FUTURE OF VIGILANT INFORMATION SYSTEMS

Western Digital's CIO, dashboard teams, and line managers continue to develop the VIS vision, enhance system functionality, and integrate the systems with management practices at WD.

Internally, output to wireless devices is planned, as are dashboards for other business processes.

Externally, WD plans to build dashboards for collaborating with supply chain partners. Such links will deepen visibility into the supply chain and thereby widen and accelerate the information inputs into the OODA loops, permitting even better synchronization along the chain. These dashboards will also give suppliers and customers visibility into WD's manufacturing process. Ultimately, the goal is to speed up the entire supply chain's OODA loops.

Figure 9 shows four types of real-time management dashboards, based on the amount of "business horse-power" required of their underlying VIS and their roots (their DNA). Business horsepower means the ability of the systems to link management decisions to operations and span multiple processes and multiple enterprises. The four are:

- *EIS Business Performance Dashboard*: Monitors business metrics and key performance indicators. Its roots are in Executive Support Systems.14 The corporate dashboard at WD is of this type.

- *Operations Control Dashboard*: Is tightly coupled to operational processes, whether in manufacturing or in service operations. Its roots are in industrial engineering and total quality management. The factory dashboard at WD is of this type.

- *Business Process Dashboard*: Is most commonly used for transactional processes. It monitors business processes across an enterprise while they are executing and can proactively reroute processes and reallocate resources. Its roots are in workflow automation. WD is moving in this direction.

- *Collaborative Dashboard*: Is shared by multiple partners in a business-to-business (B2B) supply chain and provides OODA loop synchronization across enterprises. WD is moving in this direction.

Ultimately, it's the combination of these dashboards at various enterprise levels, and their integration into management processes, that will yield the quickest and most effective OODA loops.

As the concept of the real-time enterprise develops, there will be much activity around development of supporting IT architectures and new types of dashboards. WD stitched together its own sense-and-respond IT architecture using generic IT infrastructure components. However, a new class of specialized IT architectures is evolving, specifically geared to real-time response. Examples include event-based computing architectures,[15] "publish and subscribe" architectures,[16] and B2B integration architectures. All enable seamless integration of business processes, permitting faster response.

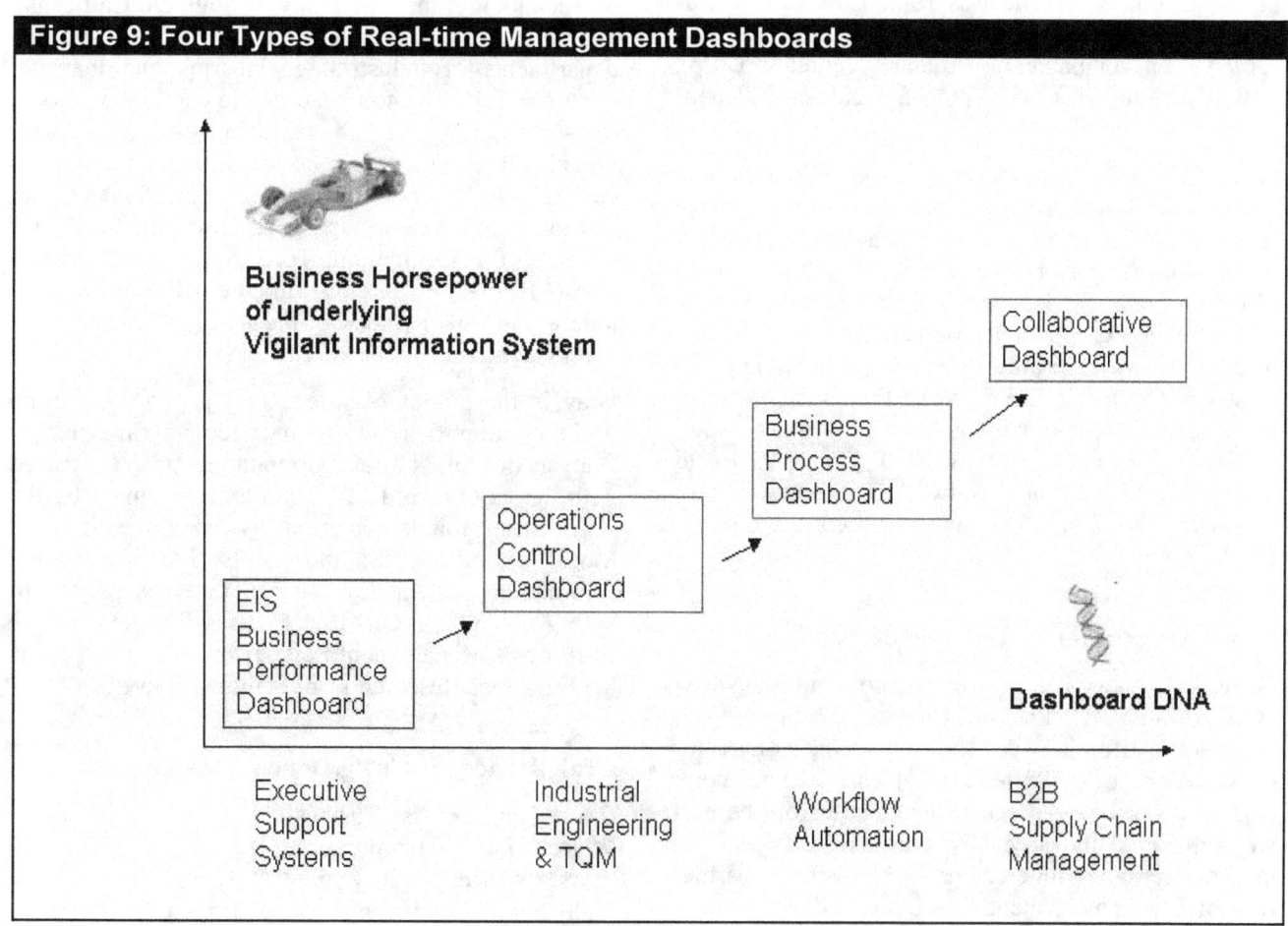

Figure 9: Four Types of Real-time Management Dashboards

ABOUT THE AUTHORS

Robert Houghton (bob.houghton@wdc.com)

Robert J. Houghton returned to Western Digital in 2000 as chief information officer and vice president, information technology. With more than 26 years in the IT field, Houghton previously served Western Digital from 1998 to 1999 as director of information services.

Reporting to Matt Massengill, chairman and chief executive officer, Houghton is responsible for all aspects of Western Digital's information technology infrastructure worldwide, including information services, network infrastructure, and technical operations, to provide a best-of-class information environment and help drive the company's expanding business objectives. As a world-class manufacturer, Western Digital's IT infrastructure includes four high-volume manufacturing facilities worldwide.

Houghton contributes to Western Digital extensive experience in operations, materials, and information systems management. Previously, he was CIO and vice president of MIS and network operation centers at Zland.com, where he was responsible for worldwide and co-location hosting data centers. He also served such companies as Adaptec, Jennings Corp., RWD Technologies and Litton Systems in IT management functions.

Houghton earned his bachelor's degree in Criminal Justice and Sociology from the University of Maryland, and information systems management certifications from the University of Maryland and University of California at Davis, as well as under the American Management Association program. He currently is a member of many professional groups, including the Southern California Chapter of SIM (Society of Information Managers); the Southern California Chapter of CIO's, Orange County; WINPOPRO.NET (World Information Professional Network, CIO's); Advisory Board for Proof Point Ventures; and Oracle Leaders Circle.

Omar El Sawy (omar.elsawy@marshall.usc.edu)

Omar A. El Sawy is Professor of Information Systems at the Marshall School of Business, University of Southern California, where he also serves as Director of Research at the Center for Telecom Management, an industry-sponsored center that focuses on the networked digital industry. His interests include redesigning electronic value chains for e-business, partner relationship management, and knowledge management and vigilance in fast-response environments. El Sawy holds a Ph.D. from Stanford Business School, an MBA from the American University in Cairo, and a BSEE from Cairo University. Prior to joining USC in 1983, he worked as an engineer and manager for twelve years, first at NCR Corporation, and then as a manager of computer services at Stanford University. He has lectured, consulted, and carried out research in four continents. El Sawy is the author of over 70 papers, and his writings have appeared in both information systems and management journals. He is the author of the book *Redesigning Enterprise Processes for e-Business*. He serves on six journal editorial boards and is a five-time winner of SIM's Paper Awards Competition.

Paul Gray (paul.gray@cgu.edu)

Paul Gray is Professor Emeritus and Founding Chair of the School of Information Science at Claremont Graduate University. His current interest in information systems include, business intelligence, knowledge management, data warehousing and electronic commerce. Before coming to Claremont in 1983, he was a professor at Stanford University, the Georgia Institute of Technology, the University of Southern California, and Southern Methodist University where he taught in departments of industrial engineering and management science. Prior to his academic career, he worked for 18 years in research and development organizations including, nine years at SRI International. He is currently a Visiting Professor at the University of California at Irvine and is affiliated with its Center for Research on Information, Technology and Organization (CRITO). He is editor-in-chief of the electronic journal *Communications of the Association for Information Systems*.

Gray is the author of three "first papers": in group decision support systems, in telecommuting, and in analysis of crime in transportation. He was recognized with the LEO award for lifetime achievement by the Association for Information Systems; the Kimball Medal of INFORMS; and the EDSIG Outstanding Information Systems Educator 2000. He is a fellow of both AIS and INFORMS. He was President of the Institute of Management Science in 1992. His PhD is in Operations Research from Stanford University.

Craig Donegan (Craig.Donegan@wdc.com)

Craig Donegan is Business Solutions Manager at Western Digital Corporation. The majority of his 30 years' experience (20 years at Western Digital) in the high-tech industry has been dedicated to Inventory management, Supply Chain optimization and Data warehousing. Craig's work experience has been divided equally between Business Operations & Information Technology. Prior to joining Western Digital Corporation, he was Senior MRP Project Manager at Comdial Corporation.

Ashish Joshi (Ashishi.Joshi@wdc.com)

Ashish has over nine years of experience in Information Technology. He currently serves as Manager of Business Applications for Western Digital Corporation. Ashish has been with Western Digital since 1997 often working in the factories and developing systems for production use.

CONTINENTAL AIRLINES FLIES HIGH WITH REAL-TIME BUSINESS INTELLIGENCE[1,2]

Ron Anderson-Lehman
Continental Airlines

Hugh J. Watson
University of Georgia

Barbara H. Wixom
University of Virginia

Jeffrey A. Hoffer
University of Dayton

Executive Summary

Real-time data warehousing and business intelligence (BI), supporting an aggressive Go Forward business plan, have helped Continental Airlines transform its industry position from "worst to first" and then from "first to favorite." With a $30M investment in hardware and software over six years, Continental has realized conservatively over $500M in increased revenues and cost savings in areas such as marketing, fraud detection, demand forecasting and tracking, and improved data center management.

Continental is now recognized as a leader in real-time business intelligence based upon its scalable and extensible architecture, prudent decisions on what data are captured in real-time, strong relationships with end users, a small and highly-competent data warehouse staff, a careful balance of strategic and tactical decision-support requirements, its understanding of the synergies between decision support and operations, and changed business processes that utilize real-time data.

CONTINENTAL TRANSFORMS ITSELF

Real-time business intelligence (BI) is taking Continental Airlines to new heights. Powered by a real-time data warehouse, the company has dramatically changed all aspects of its business. Continental's president and COO, Larry Kellner, describes the impact of real-time BI in the following way: "Real-time BI is critical to the accomplishment of our business strategy and has created significant business benefits." In fact, Continental has realized more than $500 million in cost savings and revenue generation over the past six years from its BI initiatives, producing an ROI of more than 1,000 percent.

Continental's current position is dramatically different from only ten years ago. The story begins with the arrival of Gordon Bethune as CEO, who led Continental from its "worst to first" position in the airline industry. A key to this turnaround was the Go Forward Plan, which continues to be Continental's blueprint for success and is increasingly supported by real-time BI and data warehousing.[3] Currently, the use of real-time technologies has been critical for Continental in moving from "first to favorite" among its customers, especially among its best customers.

Continental's real-time warehouse provides a powerful platform for quickly developing and deploying applications in revenue management, customer relationship management, flight and ground operations, fraud detection, security, and others. Some of these applications, the quantifiable benefits they are generating, and the technology in place that supports them are described. Continental's experiences with real-time BI and data warehousing have resulted in insights

[1] The senior accepting editor is Jack Rockart.
[2] We thank Anne Marie Reynolds, Luisa Chong, Saleem Hussaini, Carlos Ibarra, and the rest of the data warehousing team at Continental Airlines for their contributions to this article. Teradata, a division of NCR, provided funding for this case study.

[3] Some people prefer the term "right-time" over real-time in order to emphasize that data only needs to be as fresh as the decisions or business processes require. Depending on the business need, data can be hourly, daily, and even weekly or monthly and still be real-time. We use the terms real-time and right-time synonymously.

and practices from which other companies can benefit, and these lessons learned are discussed.

Decision support has evolved over the years, and the work at Continental exemplifies current practices. The article concludes by putting Continental's real-time BI and data warehousing initiatives into a larger decision-support context.

CONTINENTAL'S HISTORY

Continental Airlines was founded in 1934 with a single-engine Lockheed aircraft on dusty runways in the American Southwest. [4] Over the years, Continental has grown and successfully weathered the storms associated with the highly volatile, competitive airline industry. With headquarters in Houston, Texas, Continental is currently the USA's fifth largest airline and the seventh largest in the world. It carries approximately 50 million passengers a year to five continents (North and South America, Europe, Asia, and Australia), with over 2,300 daily departures to more than 227 destinations. Continental, along with Continental Express and Continental Connection, now serves more destinations than any other airline in the world. Numerous awards attest to its success as an airline and as a company (see Appendix A).

An Airline in Trouble

Only ten years ago, Continental was in trouble. There were ten major US airlines, and Continental ranked tenth in on-time performance, mishandled baggage, customer complaints, and denied boardings because of overbooking. Not surprisingly, with this kind of service, Continental was in financial trouble. It had filed for Chapter 11 bankruptcy protection twice in the previous ten years and was heading for a third, and likely final, bankruptcy. It had also gone through ten CEOs in ten years. People joked that Continental was a "Perfect 10." [5]

Enter Gordon Bethune and the Go Forward Plan

The rebirth of Continental began in 1994 when Gordon Bethune took the controls as CEO. He and Greg Brenneman, who was a Continental consultant at the time, conceived and sold to the Board of Directors the Go Forward Plan. It had four interrelated parts that had to be executed simultaneously.

- *Fly to Win.* Continental needed to better understand what products customers wanted and were willing to pay for.

- *Fund the Future.* It needed to change its costs and cash flow so that the airline could continue to operate.

- *Make Reliability a Reality.* It had to be an airline that got its customers to their destinations safely, on time, and with their luggage.

- *Working Together.* Continental needed to create a culture where people wanted to come to work.

Most employees supported the plan; those who did not left the company. Under Bethune's leadership, the Go Forward Plan, and a re-energized workforce, Continental made rapid strides. Within two years, it moved from "worst to first" in many airline performance metrics.

Information Wasn't Available

The movement from "worst to first" was, at first, only minimally supported by information technology. Historically, Continental had outsourced its operational systems to EDS, including the mainframe systems that provided a limited set of scheduled reports. There was no support for ad hoc queries. Each department had its own approach to data management and reporting.

The airline lacked the corporate data infrastructure for employees to quickly access the information they needed to gain key insights about the business. However, senior management's vision was to merge data into a single source, with information scattered across the organization so that employees in all departments could conduct their own business analyses to execute better and run a better and more profitable airline.

Enter Data Warehousing

This vision led to the development of an enterprise data warehouse. Janet Wejman, CIO at the time, recognized that the warehouse was a strategic project and brought the development, subsequent maintenance, and support all in-house. She believed the warehouse was core to Continental's business strategy, so it should not be outsourced. Work on the warehouse began, and after six months of development, it went into production in June 1998.

The initial focus was to provide accurate, integrated data for revenue management. Prior to the warehouse, only leg-based (a direct flight from one airport to another) data was available. Continental could therefore not track a customer's itinerary from origin to destination through several stops. Thus, Continental could

---

[4] The company history is available at www.continental/company.
[5] The story of Continental's problems and the actions that turned the company around can be read in Bethune, G. with Huler, S. *From Worst to First: Behind the Scenes of Continental's Remarkable Comeback*, Wiley, New York, 1998.

Figure 1: Three Initial Data Warehouse Applications

Demand-driven Dispatch

Prior to the warehouse, flight schedules and plane assignments were seldom changed once set, regardless of changes in markets and passenger levels. Continental operated flights without a detailed, complete understanding of each flight's contribution to profitability. After the data warehouse, Continental created a Demand-driven Dispatch application that combines forecast information from the revenue management data mart (which is integrated with the enterprise data warehouse) with flight schedule data from the data warehouse, to identify opportunities for maximizing aircraft usage. For example, the system might recommend assigning a larger plane to a flight with unusually high demand. Continental uses this application to "cherry pick" schedule changes that increase revenue. Demand-driven Dispatch has lead to an estimated $5 million dollars a year in incremental revenue.

Goodwill Letters

An eight-month test of the airline making goodwill gestures to customers showed that even small gestures can be very important to building loyalty. To make these gestures, marketing analysts used the data warehouse to marry profitability data and algorithms with customer records to identify Continental's high-value customers. The marketing department then divided these high-value customers into three groups. When any of these individuals was delayed more than 90 minutes, one group received a form letter apologizing, the second group received the letter and a free trial membership to the President's Club (a fee-based airport lounge) or some other form of compensation, and the third group received no letter at all.

Customers who received regular written communication spent 8 percent more with the airline in the next 12 months. In addition, nearly 30 percent of those receiving the President's Club trial membership joined the club following the trial, resulting in an additional $6 million in revenues. The concept of goodwill letters was expanded across the company to include the top 10 percent of Continental's customers.

Group Snoop

Group Snoop refers to a fare rule and contract compliance application that attempts to reduce the risk and financial impact of "no show" customers for any given flight. Because of the impact that groups can have on the final number of passengers boarded on a flight, advanced deposits and other contractual obligations are required for bookings of groups of 10 or more people who are traveling together.

However, travel agents can bypass this requirement and book a group of 16 by making two bookings of seven and nine without deposits or contracts. The fare rule has therefore created an incentive for agencies to block space in smaller groups to avoid making a deposit. Should the group not materialize, the financial impact to the airline can be significant. Sometimes agents convert smaller bookings to a group, but sometimes the bookings merely hold inventory space.

Using the booking and agency data from the warehouse, this Group Snoop application sorts reservations by booking agent and travel agent and then queries all groups of less than ten to identify the same travel agent ID and itinerary. Continental can then assess seat inventory more accurately and get travel agents to comply with the group booking requirements. Group Snoop has provided Continental an annualized savings of $2 million.

not study market and customer behavior, nor optimize its entire network of flights. The warehouse integrated multiple data sources – flight schedule data, customer data, inventory data, and more – to support pricing and revenue management decision-making based on journey information.

The data warehouse provided a variety of early, big "wins" for the business. The initial applications were soon followed by applications that required integrating customer information, finance, flight information, and security. These applications created significant financial lift in all areas of the Go Forward Plan. Figure 1 gives three examples of how the new integrated enterprise data was initially used at Continental.

Raising the Bar to "First to Favorite"

Once Continental achieved its goals of returning to profitability and ranking first in the airline industry in many performance metrics, Gordon Bethune and his management team raised the bar by expanding the vision. Instead of merely performing best, they wanted Continental to be their customers' favorite airline.

The First to Favorite strategy builds on Continental's operational success and focuses on creating customer loyalty by treating customers extremely well, especially the high-value customers (who are called Co-Stars). Figure 2 shows a poster in Continental's headquarters that reminds employees of the First to Favorite initiative

Figure 2: Continental Airline Internal Communications Poster

"WORST TO FIRST" SHOOK THE INDUSTRY.

"FIRST TO FAVORITE" REINVENTS IT.

Now you have a way to recognize CoStars. Offer them more personalized service and help make Continental their favorite airline.

Continental Airlines

Work Hard. Fly Right.

The Go Forward Plan identified more actionable ways the company could move from first to favorite. Technology became increasingly critical for supporting these new initiatives. At first, having access to historical, integrated information was sufficient for the Go Forward Plan, and it generated considerable strategic value for the company. However, as Continental moved ahead with the First to Favorite strategy, it became increasingly important for the data warehouse to provide real-time, actionable information to support enterprise-wide tactical decision-making and business processes.

Fortunately, the warehouse team had anticipated and prepared for the ultimate move to real-time.[6] From the outset of the warehouse project, they built an architecture able to handle real-time data feeds into the warehouse, extracts of data from the warehouse into legacy systems, and tactical queries to the warehouse that required sub-second response times.[7] In 2001, real-time data became available in the warehouse.

REAL-TIME BI APPLICATIONS

The amount of real-time data in the warehouse grew quickly. Continental moves real-time data (ranging from to-the-minute to hourly) about customers, reservations, check-ins, operations, and flights from its main operational systems to the warehouse. The following sections illustrate the variety of key applications that rely on real-time data. Many of the applications also use historical data from the warehouse.

Fare Design

To offer competitive prices for flights to desired places at convenient times, Continental uses real-time data to optimize airfares (using mathematical programming models). Once a change is made in price, revenue management immediately begins tracking the impact of that price on future bookings. Knowing immediately how a fare is selling allows the group to adjust how many seats should be sold at each given price. Last-minute customized discounts can be offered to the most profitable customers to bring in new revenue, as well as to increase customer satisfaction. Continental has earned an estimated $10 million annually through fare design activities.

Recovering Lost Airline Reservations

In 2002, an error in Continental's reservation system resulted in a loss of 60,000 reservations. Within a matter of hours, the warehouse team developed an application whereby agents could obtain a customer's itinerary and confirm whether the passenger was booked on flights.

Another similar situation happened in 2004 when the reservation system had problems communicating with other airlines' reservation systems. In certain circumstances, the system was not sending reservation information to other airlines, and, consequently, other airlines weren't reserving seats for Continental's passengers. As a result, Continental customers would arrive for a flight and not have a seat. Once the problem was discovered, the data warehouse team was able to run a query to get the information on passengers who were affected but who had not yet flown.

---

[6] An excellent research report on real-time BI is White, C. *Building the Real-Time Enterprise*," The Data Warehousing Institute, Seattle, WA, 2003.

[7] Insights about the methods and challenges of providing real-time data feeds is provided in Brobst, S. "Delivery of Extreme Data Freshness with Active Data Warehousing," *Journal of Data Warehousing* (7:2), Spring 2002, pp.4-9.

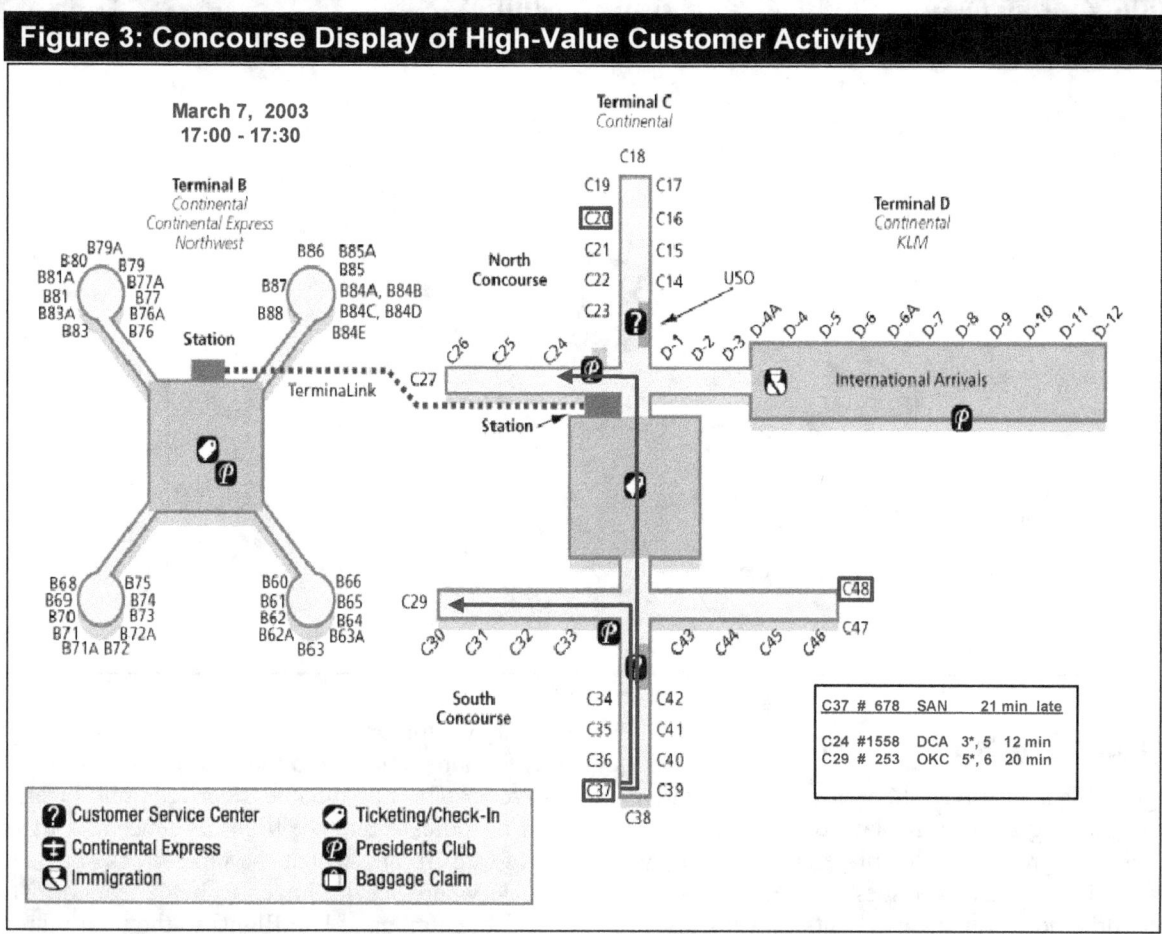

Figure 3: Concourse Display of High-Value Customer Activity

This information was fed back into the reservation system so that seats could be assigned, thus avoiding a serious customer relations problem.

Customer Value Analysis

A customer value model using frequency, recency, and monetary value gives Continental an understanding of its most profitable customers. Every month, the customer value analysis is performed using data in the data warehouse, and the value is fed to Continental's customer-facing systems so that employees across the airline, regardless of department, can recognize their best customers when interacting with them.

This knowledge helps Continental react quickly, effectively, and intelligently in tough situations. For example, just after 9/11, Continental used customer value information to understand where its best customers were stranded around the world. Continental applied this information to its flight scheduling priorities, and, while the schedules were being revised, the company worked with its lodging and rental car partners to make arrangements for its stranded customers. The highest value customer was in Zurich, and he used Continental's offices to conduct business until he was able to fly home.

Marketing Insight

Marketing Insight was developed to provide sales personnel, marketing managers, and flight personnel (e.g., ticket agents, gate agents, flight attendants, and international concierges) with customer profiles. This information, which includes seating preferences, recent flight disruptions, service history, and customer value, is used to personalize interactions with customers.

Gate agents, for instance, can pull up customer information on their screen and drill into flight history to see which high-value customers have had flight disruptions. Flight attendants receive this information on their "final report," which lists the passengers on their flights, including customer value information. A commonly told story is about a flight attendant who learned from the final report that one of the high-value customers on board recently experienced a serious delay. She apologized to the customer and thanked him for his continuing business. The passenger was very suprised that she knew about the incident and cared enough to apologize.

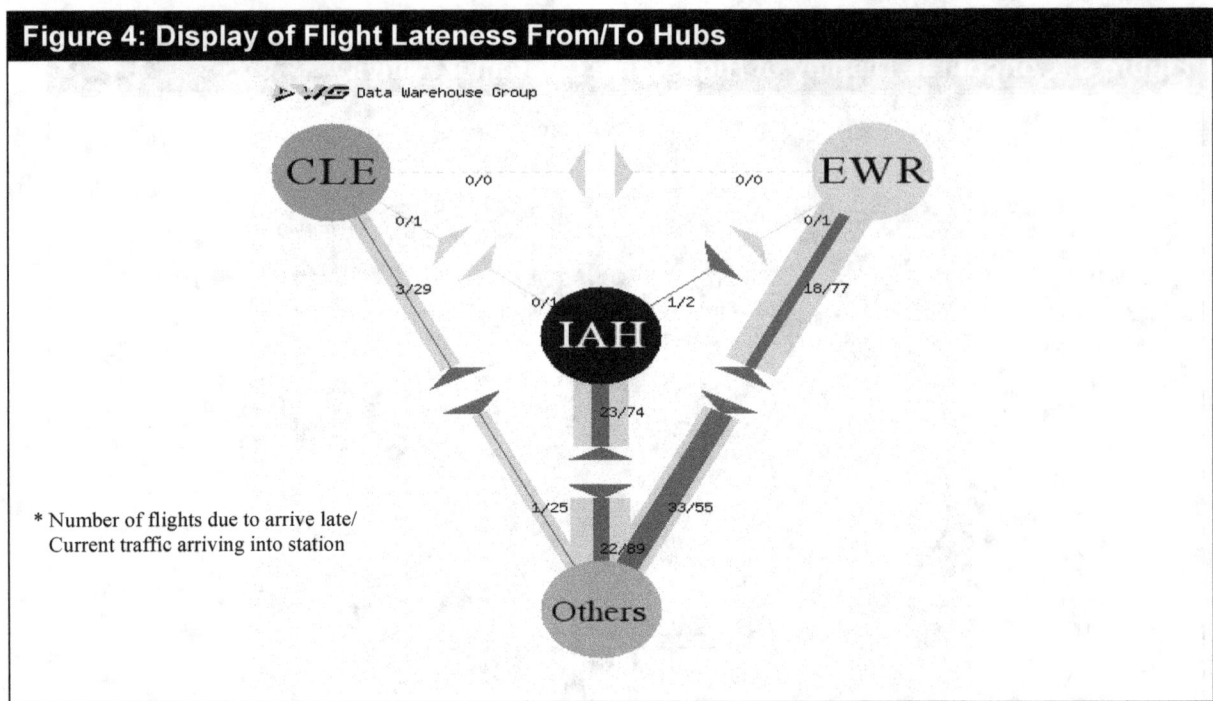

Figure 4: Display of Flight Lateness From/To Hubs

Flight Management Dashboard

The Flight Management Dashboard is an innovative set of interactive graphical displays developed by the data warehouse group. The displays help the operations staff quickly identify issues in the Continental flight network and then manage flights in ways to improve customer satisfaction and airline profitability.

Some of the dashboard's displays help Operations better serve Continental's high-value customers. For example, one display, a graphical depiction of a concourse, is used to assess where Continental's high-value customers with potential service issues are located or will be in a particular airport hub (see Figure 3). The display shows where these customers have potential gate-connection problems so that gate agents, baggage supervisors, and other operations managers can provide ground transportation assistance and other services so that these customers and their luggage do not miss flights.

Figure 3 shows that Flight 678 is arriving 21 minutes late to Gate C37 and eight high-value customers need assistance in making their connections to Gates C24 (three passengers) and C29 (five passangers), and they have 12 minutes and 20 minutes, respectively, to catch their flights.

On-time arrival is an important operational measure at Continental. The Federal Aviation Administration requires airlines to report arrival times and provide the summary statistics to the flying public. Therefore, another critical set of dashboard displays helps Operations keep arrivals and departures on time. One dis-

play shows the traffic volume between the three Continental hub stations and the rest of their network (see Figure 4). The line thickness between hub locations is used to indicate relative flight volumes and the number of late flights so that the Operations staff can anticipate where services need to be expedited. The ratio of the number of late flights to the total number of flights between the hubs also is shown. The Operations staff can drill down to see individual flight information by clicking on the lines between the hub locations.

Another line graph summarizes flight lateness. Users can drill down to detailed pie charts that show degrees of lateness, and within each pie, to the individual flights in that category. Still another chart concentrates on flights between the US and Europe and the Caribbean. It can show similar critical flight statistics.

In all of these elements of the dashboard, high-level views can be broken down to show the details on customers or flights that compose different statistics or categories.

Fraud Investigations

In the wake of 9/11, Continental realized that it had the technology and data to monitor passenger reservation and flight manifests in real-time. A "prowler application" was built so that corporate security could search for names or patterns. More than 100 "profiles" are run regularly against the data to proactively find fraudulent activity. When matches are found, e-mail and a message page are sent immediately to corporate security. Not only does this application allow corporate security to prevent fraud, but it also enhances their ability to gather critical intelligence through more timely interviews with suspects, victims, and witnesses.

One profile, for example, looks for reservations agents who make an extraordinary number of first-class bookings. Last year, Continental was able to convict an agent who was manufacturing false tickets and then exchanging them for real first-class tickets that she sold to her friends. Continental received over $200,000 in restitution from that one case. In total, Continental was able to identify and prevent more than $15 million in fraud in 2003 alone.

Is it Safe to Fly

Immediately after the terrorist attacks of 9/11, planes were ordered to land at the nearest airport. Continental had 95 planes that did not reach their planned destination. Sometimes there were three or four planes at a little airport in a town with no hotels, and passengers had to move in with the local people. At Continental's headquarters, FBI agents moved into a conference room with a list of people they had authority to check. Queries were run against flight manifest data to see if potential terrorists were on flights, and it was only after a flight was deemed safe that it was allowed to fly. Continental Airlines was recognized by the FBI for its assistance in the investigations in connection with 9/11.

SUPPORTING FIRST TO FAVORITE WITH TECHNOLOGY

Real-time BI requires appropriate technologies that is, those that extend traditional BI and data warehousing. At Continental, real-time technologies, and the associated processes, are critical for supporting the First to Favorite strategy.

The Data Warehouse

Continental's real-time BI initiative is built on the foundation of an 8-terabyte enterprise Teradata Warehouse running on a 3 GHz, 10-node NCR 5380 server.[8] The data warehouse supports 1,292 users who access 42 subject areas, 35 data marts, and 29 applications. Figure 5 shows the growth of the data warehouse over time.

The basic architecture of the data warehouse is shown in Figure 6. Data from 25 internal operational systems (e.g., the reservations system) and two external data sources (e.g., standard airport codes) are loaded into

---

[8] Teradata uses the "active" data warehousing term to describe real-time data warehousing.

the data warehouse. Some of these sources are loaded in real-time and others in batch, based on the capabilities of the source and the business need. Critical information determined from analyses in the data warehouse (e.g., customer value) is fed from the data warehouse back into the operational systems.

Figure 5: Warehouse Growth Over Time			
	1998	2001	Current
Users	45	968	1292
Tables	754	5851	16226
Subject Areas	11	33	42
Data Marts	2	23	35
Applications	0	12	29
DW Personnel	9	15	15

Data Access

Users access the warehouse data in various ways (see Figure 7). Some use standard query interfaces and analysis tools, such as Teradata's QueryMan, Microsoft Excel, and Microsoft Access. Others use custom-built applications. Still others use either the desktop (i.e., "fat client") or Web versions of Hyperion Intelligence. An estimated 500 reports have been created in Hyperion Intelligence, and many of these reports are pushed to users at scheduled intervals (e.g., at the first of the month, after the general ledger is closed). Other products include SAS' Clementine for data mining and Teradata CRM for campaign management.

Real-time Data Sources

The data warehouse's real-time data sources range from the mainframe reservation system to satellite feeds transmitted from airplanes to a central customer database. Some data feeds are pulled from the sources in batch mode. For example, files of reservation data are extracted and sent using FTP (file transport protocol) from a mainframe application on an hourly basis. An application converts the data into 3rd normal form and writes the updated records to queues for immediate loading into the data warehouse.

Other data feeds are loaded to the warehouse within seconds. Flight data (called FSIR, or flight system information record) is sent real-time from airplanes

---

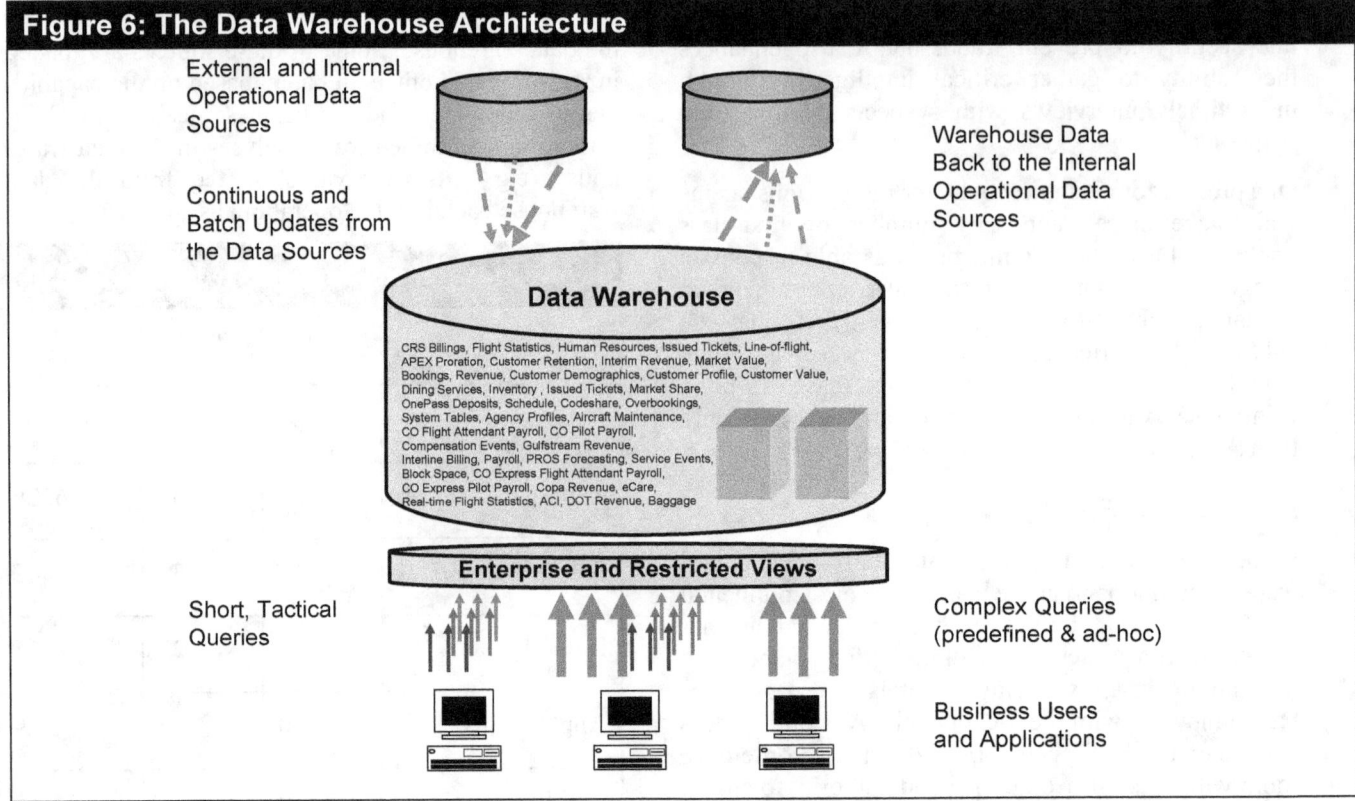

Figure 6: The Data Warehouse Architecture

External and Internal Operational Data Sources

Continuous and Batch Updates from the Data Sources

Warehouse Data Back to the Internal Operational Data Sources

Data Warehouse

CRS Billings, Flight Statistics, Human Resources, Issued Tickets, Line-of-flight, APEX Proration, Customer Retention, Interim Revenue, Market Value, Bookings, Revenue, Customer Demographics, Customer Profile, Customer Value, Dining Services, Inventory , Issued Tickets, Market Share, OnePass Deposits, Schedule, Codeshare, Overbookings, System Tables, Agency Profiles, Aircraft Maintenance, CO Flight Attendant Payroll, CO Pilot Payroll, Compensation Events, Gulfstream Revenue, Interline Billing, Payroll, PROS Forecasting, Service Events, Block Space, CO Express Flight Attendant Payroll, CO Express Pilot Payroll, Copa Revenue, eCare, Real-time Flight Statistics, ACI, DOT Revenue, Baggage

Enterprise and Restricted Views

Short, Tactical Queries

Complex Queries (predefined & ad-hoc)

Business Users and Applications

via satellite to an operations control center system. FSIR data may include time estimates for arrival, the exact time of lift-off, aircraft speed, etc. This data is captured by a special computer and placed in a data warehouse queue, which is then immediately loaded into the warehouse.

Other data sources are pushed real-time by the sources themselves, triggered by events. For example, Continental's reservations system, OnePass frequent flier program, Continental.com, and customer service applications all directly update a central customer database. Every change made to a customer record in the customer database activates a trigger that pushes the update as XML encoded data to a queue for immediate loading into the data warehouse.

The Data Warehouse Team

The data warehouse team has 15 people who are responsible for managing the warehouse; developing and maintaining the infrastructure; data modeling; developing and maintaining data extraction, transformation and loading processes; and working with the business units. The organization chart for the data warehouse staff is shown in Figure 8.

Data Warehouse Governance

The Data Warehouse Steering Committee provides direction and guidance for the data warehouse. This large, senior-level committee has 30 members, most at the Director level and above. They come from the business areas supported by the data warehouse and are the spokespersons for their areas. Business areas that intend to participate in the warehouse are invited to join the committee. The warehouse staff meets with the committee to inform and educate the members about warehouse-related issues. In turn, the members identify business-area opportunities for the warehouse staff. They also help the warehouse team justify and write requests for additional funding. Another responsibility is to help set priorities for future directions for the data warehouse.

Securing Funding

The business areas drive the funding for the data warehouse. There has always been one area that has helped either justify the initial development of the warehouse or encourage its later expansion. Revenue Management supported the original development. The second and third expansions were justified by Marketing to support the Worst to First, and then First to Favorite strategies. Corporate Security championed the fourth, and most recent, expansion. This approach to funding helps ensure that the data warehouse supports the needs of the business.

Figure 7: Data Warehouse Access

Application or Tool	Types of Users	Number of Users
Hyperion Intelligence – Quickview (web)	Enterprise	300
Hyperion Intelligence – Explorer (desktop)	Enterprise	114
Access	Enterprise	200
Custom Applications	Enterprise	700
Teradata CRM	Marketing	20
Clementine Data Mining	Revenue Management	10
Teradata QueryMan	Enterprise	150
Excel	Enterprise	Many

The funding does not come directly from the business areas (i.e., their budgets). Rather, the funding process treats proposals as a separate capital expense. However, the business areas must supply the anticipated benefits for the proposals. Therefore, any proposal must have a business partner who identifies and stands behind the benefits.

THE BENEFITS OF BUSINESS INTELLIGENCE

Continental has invested approximately $30 million into real-time warehousing over the past six years. Of this amount, $20 million was for hardware and software, and $10 million for personnel. Although this investment is significant, the quantifiable benefits are magnitudes larger. Specifically, over the past six years, Continental has realized over $500 million in increased revenues and cost savings, resulting in a ROI of over 1,000 percent.

The benefits range from better pricing of tickets to increased travel to fraud detection. Figure 9 identifies some realized benefits. Because almost 1,300 users have warehouse access, it is impossible to know all the benefits. However, when big "wins" are achieved, the benefits are recorded and communicated throughout the company. This internal publicity helps preserve the excitement around warehouse use, and encourages business users to support warehouse expansion efforts.

LESSONS LEARNED

The experiences at Continental confirm the commonly known keys to success for any enterprise-wide IT initiative: the need for senior management sponsorship and involvement, close alignment between business and IT strategies, a careful selection of technologies, ongoing communication, a clear vision and roadmap, and letting the business drive the technology. More interesting, though, are the following seven insights

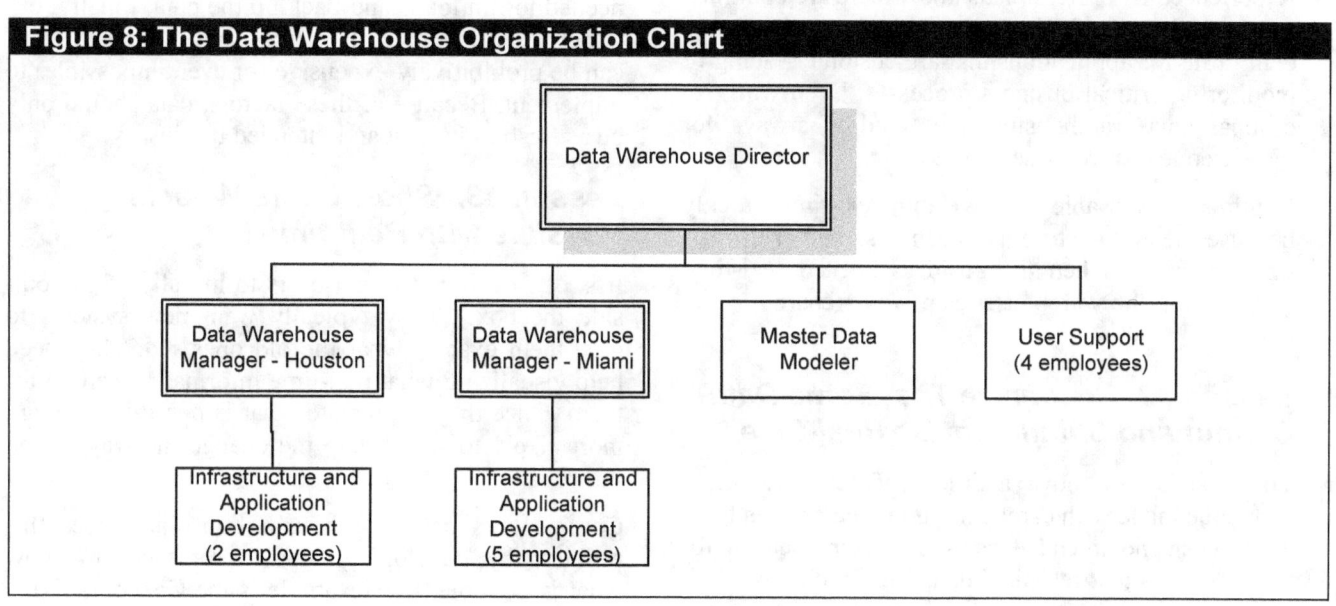

Figure 8: The Data Warehouse Organization Chart

Figure 9: Sample Benefits from Real-time BI and Data Warehousing

Marketing	• Continental performs customer segmentation, target marketing, loyalty/retention management, customer acquisition, channel optimization, and campaign management using the data warehouse. Targeted promotions have produced cost savings and incremental revenue of $15 to $18 million per year. • A targeted CRM program resulted in $150 million in additional revenues in one year, while the rest of the airline industry declined 5 percent. • Over the past year, a goal was to increase the amount of travel by Continental's most valuable customers. There has been an average increase in travel of $800 for each of the top 35,000 customers.
Corporate Security	• Continental was able to identify and prevent over $30 million in fraud over the past three years. This prevention resulted in more than $7 million in cash collected.
IT	• The warehouse technology has significantly improved data center management, leading to cost savings of $20 million in capital and $15 million in recurring data center costs.
Revenue Management	• Tracking and forecasting demand has resulted in $5 million in incremental revenue. • Fare design and analysis improves the ability to gauge the impact of fare sales, and these activities have been estimated to earn $10 million annually. • Full reservation analysis has realized $20 million in savings through alliances, overbooking systems, and demand-based scheduling.

learned especially about the development and implementation of real-time BI.

Lesson #1: Prepare Early On for Real-time BI

Experienced BI professionals know there are continual demands for ever-fresher data. This demand is especially true for applications that are customer-facing or monitoring critical business processes. Even with traditional data warehousing, the trend is always for more frequent warehouse updates.

Continental was able to move into real-time quickly because the architecture had been designed with real-time in mind. When the business needed real-time information, the warehouse team was prepared to deliver it.

Lesson #2: Recognize That Some Data Cannot and Should not Be Real-time

The decision to move additional data to real-time should be made with care; data should be only as fresh as its cost and intended use justify. One reason for taking care is that real-time data feeds are more difficult to manage. The real-time processes, such as the

flow of transaction data into queues, must be monitored constantly because problems can occur throughout the day (rather than just when a batch update is run). And, when problems with data occur, they must be addressed immediately, putting pressure on staffing requirements. Also, additional hardware may be needed to run loads and back up the data. Finally, obtaining real-time data feeds from some source systems can be prohibitively expensive (or even impossible) to implement. Because of these factors, data should only be as fresh as its cost and intended use justify.

Lesson #3: Show Users What Is Possible with Real-time BI

It is often difficult to get users to initially "think outside the box." They typically want new systems to give them exactly what the old ones did. They need help visualizing what real-time information can do for them. Once they appreciate what is possible, they are more likely to say: "Help me change the way we do business."

Continental's data warehousing staff addresses this problem by developing "cool" prototypes to show what is possible. One example is the Concourse Map application described earlier. When the users saw how data could be depicted in graphical ways (e.g., as an actual concourse with colors and lines with special meaning), they came up with their own ideas for how real-time data could help them operate the hubs better. At Continental, the current challenge is to find the time to support the many new ideas the users have.

Lesson #4: Adjust the Skill Mix on Both the Warehouse and Business Sides

In many companies, the data warehousing staff generally has strong technical skills but limited business knowledge, while the business side has limited technical skills but good business knowledge. At the intersection of the warehousing and business organizations, there is a dramatic change in the technical/business skills and knowledge mix.

At Continental, this change is very gradual across the warehouse/ business intersection. Those warehouse personnel who work closest with the business users have considerable business knowledge. On the other hand, many business users have developed excellent technical skills – in fact, enough knowledge to build their own warehouse applications. The gradual shift in skills has reduced what can be a significant "divide," and helps ensure that Continental's warehouse is used to support the business.

Lesson #5: Manage Strategic and Tactical Decision Support to Co-exist

Strategic and tactical decision support have different characteristics, yet they must co-exist in the same warehouse environment. Strategic decision support typically involves the analysis of large amounts of data that must be "sliced and diced" in various ways. Tactical (sometimes called "operational") decision support often requires repeatedly accessing and analyzing only a limited amount of data with a sub-second response time.

Successful support for both requires both business and technical solutions. On the business side, priorities must be set for the processing of queries from users and applications. For example, a tactical query should have a higher priority than a strategic data-mining application. On the technical side, a query manager must recognize priorities, monitor queries, defer long-running queries for later execution, and dynamically allocate query resources.

At Continental, tactical queries that access single records are set to high priority. These queries usually come from applications, such as continental.com, that require instantaneous response time. All daytime batch data loads are set to low priority, and all day-

time trickle feed loads are set to medium priority. And, users who perform ad-hoc queries are given medium priority access.

Lesson #6: Real-time BI Blurs the Line Between Decision Support and Operational Systems

For one thing, the performance requirements for real-time BI (e.g., response time, downtime) are similar to those of operational systems. In fact, the same personnel (or ones with similar skills) may be used for both. Whereas decision support and operational systems may previously have had their own standards, because of the need for closer system integration, common standards become more important. In nearly all instances, the warehouse needs to be compatible with overall IT standards. Furthermore, the data warehouse team must be made aware of any upcoming changes to any operational system that provides real-time data because these changes could have immediate and potentially disruptive impacts.

Lesson #7: Real-time BI Doesn't Deliver Value Unless Downstream Decision-making and Business Processes Are Changed

There are three sources of latency in real-time BI: the time to extract data from source systems, the time to analyze the data, and the time to act upon the data. The first two can be minimized using real-time technologies. The third requires getting people and processes to change. Unless downstream decision-making and business processes are changed to utilize real-time data, the value of the data decreases exponentially with the passage of time.

PUTTING THE WORK AT CONTINENTAL IN A DECISION SUPPORT PERSPECTIVE

The initial thinking, research, and practice of computer-based decision support started in the late 1960s. Prior to then, computers were used almost exclusively for transaction processing (with the exception of scientific applications). Books by Scott Morton and Keen and Scott Morton helped to create awareness of the potential of computer-based decision support.[9] Decision support systems (DSS) was the name given to

---

[9] Scott Morton, M. *Management Decision Systems: Computer-Based Support for Decision Making*, Division of Research, Harvard, Cambridge, MA, 1971. Keen, P.G.W. and Scott Morton, M. *Decision Support Systems: An Organizational Perspective*, Addison-Wesley, Reading, MA, 1978.

this type of application, and it continues in academia to be both the name of a discipline and a specific type of application. Throughout the 1970s and into the 1980s, DSS was the "hot topic" in both academia and practice. The Sprague and Carlson book (published in 1982) codified much of what had been learned about DSS, including the need for a dedicated decision-support database.[10]

In the early 1980s, the decision support focus turned to executive information systems (EIS). Rockart and Treacy's article, "The CEO Goes Online" published in the *Harvard Business Review*, and Rockart and DeLong's book, *Executive Support Systems: The Emergence of Top Management Computer Use*, did much to publicize and create interest in EIS.[11] The research and work on EIS provided many insights that have influenced current practice. For example, many EIS failures were related to an inadequate data infrastructure, which supported the emergence of data warehouses.[12] The use of critical success factors is now seen in business performance management (BPM), digital dashboards, and balanced scorecards, all popular today.[13] In successful decision-support applications, there are continuing pressures to provide users with ever-fresher data.[14]

In the late 1980s, data warehousing emerged to provide the data infrastructure needed for decision-support applications. The writings of Inmon (who is widely recognized as "the father of data warehousing") and Kimball helped many organizations think about and develop data warehouses.[15] Among the first warehouse adopters were firms in telecommunications, retail, and financial services, which are highly competitive and need to understand and use customer data to be competitive.

Initially, data warehouses were perceived as a repository of historical data and were used primarily to support strategic decision making. As companies recognized the need and potential to support tactical and operational decisions, they developed operational data stores (ODS) to meet the need for very fresh data.[16] For the warehouse itself, extraction-transformation-loading (ETL) processes became more frequent, once again to provide more current data for decision support. Recognizing the need for real-time data, in the past couple of years, vendors have introduced products that allow companies to move to real-time warehousing. As companies make this move, the distinctions between operational and decision-support systems blur.[17]

Data warehouses differ in the ways companies use them. In some cases, warehouses primarily support reporting and queries, while in others they provide critical support for applications that are aligned with a company's business strategy.[18] When companies move to real-time warehousing and BI, they are better able to support tactical and operational decisions. This is both a natural evolution and a dramatic shift. It is natural in terms of the movement to providing ever-fresher data, but is a significant change in how the data can be used. With current data, it is possible to support many additional kinds of decisions and use the data to support internal operations and interactions with customers. For example, business activity monitoring (BAM) is dependent on the availability of up-to-date data.[19]

As decision support has evolved over the years, a new term emerged in the industry for analytical applications. In the early 1990s, the influential Gartner Group began to use the term business intelligence (BI), and it is now well entrenched. BI applications include DSS, on-line analytical processing (OLAP), EIS, and data mining. The BI term is only now beginning to find its way into academia's vocabulary.

Continental Airlines provides an outstanding example of how decision support is changing in many leading companies. Real-time data warehousing and BI allow Continental to use extremely fresh data to support current decision making and business processes to affect the organization's fate. Continental has put in place a decision-support infrastructure that is able to evolve with the needs of the business. Organizations must understand the natural evolution of decision support –

[10] Sprague, R.H., Jr. and Carlson, E.D. *Building Effective Decision Support Systems*, Prentice-Hall, Englewood Cliffs, NJ, 1982, pp. 221-255.

[11] Rockart, J.F. and Treacy, M.E. "The CEO Goes On-Line," *Harvard Business Review* (60:1), January-February 1982, pp. 81-93. Rockart, J.F. and DeLong, D.W. *Executive Support Systems: The Emergence of Top Management Computer Use*, Dow Jones-Irwin, Homewood, IL, 1988.

[12] Gray, P. and Watson, H.J. *Decision Support in the Data Warehouse*, Prentice-Hall, Upper Saddle River, NJ, 1998, pp. 4-6.

[13] Gregory, M.A. "Keys to Successful Performance Management," *Business Intelligence Journal* (9:1), Winter 2004, pp. 41-48.

[14] Paller, A. with Laska, R. *The EIS Book*, Dow Jones-Irwin, Homewood, IL, 1990, pp. 50-51.

[15] Inmon, W.H. *Building the Data Warehouse*, Wiley, New York, 1992. Kimball, R. *The Data Warehouse Toolkit: Practical Techniques for Building Dimensional Data Warehouses*, Wiley, New York, 1992.

[16] Gray and Watson, op. cit.

[17] Atre, S. and Malhotra, D. "Real-Time Analysis and Data Integration for BI," *DM Review* (14:2), February 2004, pp.18-20.

[18] For examples of the different ways that data warehouses are used, see Watson, H.J., Goodhue, D.L. and Wixom, B.J. "The Benefits of Data Warehousing: Why Some Companies Realize Exceptional Payoffs," *Information and Management* (39:6), May 2002, pp. 491-502. Goodhue, D.L., Wixom, B., and Watson, H.J., "Realizing Business Benefits Through CRM: Hitting the Right Target in the Right Way" *MISQE*, June 2002, pp.79-94

[19] White, C. "Now Is the Right Time for Real Time BI," *DM Review*, (14:9) September 2004, pp. 47, 50, 52, 54.

how far the field has come and its future possibilities – so that they, too, can be prepared to harness the power of real-time BI to make the right decisions at the right time.

ABOUT THE AUTHORS

Ron Anderson-Lehman

Ron Anderson-Lehman (rona-l@coair.com) is vice president and chief information officer of Continental Airlines Inc., a position he has held since August 2004. As vice president and CIO, Anderson-Lehman is responsible for all activities taken on by the Technology division, providing technical solutions to the various business units of the airline and overseeing a staff of over 300 employees. He joined the airline in 2000 as managing director of Technology before being named staff vice president of Technology in 2003.

Anderson-Lehman began his career in aviation in 1986 as a computer programmer for United Airlines. From there, he moved into roles of increasing responsibility at Covia and Galileo International before joining Continental.

Anderson-Lehman attended Iowa State University where he earned a Bachelor of Science degree in Computer Science with a minor in Mathematics. He currently serves on the board of directors for the OpenTravel Alliance.

Anderson-Lehman resides in the Houston area with his wife and three children.

Dr. Hugh Watson

Dr. Hugh Watson (hwatson@uga.edu) is a Professor of MIS in the Terry College of Business at the University of Georgia and a holder of a C. Herman and Mary Virginia Terry Chair of Business Administration. He is a leading scholar and authority on decision support, having authored 22 books and over 100 scholarly articles in journals such as *MIS Quarterly*, *Journal of MIS*, *Management Science*, and the *Academy of Management Journal*. Hugh helped develop the conceptual foundation for decision support systems in the 1970s, researched the development and implementation of executive information systems in the 1980s, and most recently, specializes in business intelligence and data warehousing.

Hugh is a Fellow of The Data Warehousing Institute and the Association for Information Systems and is the senior editor of the *Business Intelligence Journal*. He is also senior director of the Teradata University Network, a free portal for faculty who teach and research data warehousing, BI/DSS, and database. For the past 17 years, Watson has been the consulting editor for John Wiley & Sons' MIS series.

Dr. Barbara Wixom

Dr. Barbara Wixom (bwixom@mindspring.com) is an associate professor of Commerce at the University of Virginia's McIntire School of Commerce. She is an associate editor of *The Business Intelligence Journal*, a Fellow of The Data Warehousing Institute, and an instructor in data warehousing, database, and strategy at the undergraduate and graduate levels. Wixom won best paper awards from the Society for Information Management in 1999, 2000, and 2004 for her data warehousing case studies. In 2002, She won the University of Virginia's all-University Teaching Award, which recognizes the university's top professors. She has published in journals that include *Information Systems Research*, *MIS Quarterly*, and the *Journal of MIS* and has written two textbooks with John Wiley & Sons.

Dr. Jeffrey A. Hoffer

Dr. Jeffrey A. Hoffer (hoffer@udayton.edu) is the Sherman–Standard Register Professor of Data Management in the MIS, Operations Management, and Decision Sciences Department at the University of Dayton. He received a PhD from Cornell University in 1973 and was on the faculties of Case Western Reserve University and Indiana University before joining UD. He is a founder of the INFORMS College on Information Systems, the International Conference on Information Systems (and its conference chair in 1985), and the Association for Information Systems. He is author of many scholarly publications in the areas of database management, systems analysis, strategic systems planning, and human-computer interaction. He is co-author of several leading textbooks: *Modern Database Management*, *Modern Systems Analysis and Design*, *Essentials of Systems Analysis and Design*, *Object-Oriented Systems Analysis and Design*, and *Managing Information Technology: What Managers Need to Know*, all published by Prentice-Hall. Dr. Hoffer is also an associate director of the Teradata University Network, the leading Web portal for faculty and students in the data management, data warehousing, decision support, and business intelligence areas.

APPENDIX A: HONORS AND AWARDS

Best Customer Service	J.D. Power, *SmartMoney*, Ziff Davis *Smart Business*
Best International or Premium Class Service	OAG, National Airline Quality Rating, *Nikkei Business Magazine, Travel Trade Gazette Europa*, Inflight Research Services, *Condé Nast Traveler, Smart Money, Wall Street Journal*
Best Airline	*Fortune, Air Transport World, Investor's Business Daily, Hispanic Magazine, Aviation Week*, OAG
Best Technology	#1 Airline, #2 of 500 Companies – *InformationWeek*, #1 Web, by Forrester, Gomez Advisors, NPD New Media Services and InsideFlyer, TDWI 2003 Best Practice Award – Enterprise Data Warehouse, TDWI 2003 Leadership Award, CIO Enterprise Value Award

ADVANCED BUSINESS INTELLIGENCE AT CARDINAL HEALTH[1]

Traci A. Carte
Albert B. Schwarzkopf
Teresa M. Shaft
Robert W. Zmud

University of Oklahoma

Executive Summary

In the mid-1990s, Cardinal Health's Medical Products and Services business implemented SAP R/3, and built an accompanying data warehouse to handle business reporting. Since that time, use of the data warehouse has diffused widely across the enterprise and has evolved into an advanced business intelligence (BI) capability that the business professionals use regularly to solve problems and to take advantage of opportunities. Besides the data warehouse, the key components of this advanced BI environment are Cardinal Health's data infrastructure (its enterprise-wide data model, limited set of tools, and robust support environment) and its information culture (its data-driven decision style, business-led IT decision-making, dense social networks, and pull reporting structure). This article describes this advanced BI environment, how Cardinal Health evolved it, and how others might do the same.[2]

BUSINESS INTELLIGENCE IS EVOLVING

Most organizations have made substantial investments in information technology (IT) over the past decade to reconfigure their business and technology architectures in order to enhance their operating capabilities, enrich their employees' information environment, and enable events that take place in one part of the enterprise to be visible across the extended enterprise.[3] While most firms have generally introduced new technologies successfully, their users often have not taken full advantage of all the features: "evidence strongly suggests that organizations underutilize the functional potential of the majority of this mass of installed IT applications: users employ quite narrow feature breadths, operate at low levels of feature use, and rarely initiate technology- or task-related extensions of the available features."[4]

One active IT investment arena has involved data warehousing.[5] An increasingly common motive for implementing data warehouses is to improve a firm's business intelligence (BI), that is, the accumulated knowledge about itself, its operations, and, to a lesser extent, its immediate business environment, e.g., suppliers, customers, and competitors.[6] While BI solu-

---

[1] Jack Rockart was the accepting Senior Editor for this article.
[2] This study was supported by the University of Oklahoma Center for MIS Studies. A previous version was presented at the 2004 SIM Academic Workshop in Washington, D.C, December 10, 2004.
[3] For exemplars, see Anderson-Lehman, R., Watson, H. J., Wixom, B., and Hoffer, J. A., "Continental Airlines Flies High with Real-Time Business Intelligence," *MISQ Executive* (3:4), 2004, pp. 163-176; Lee, H., Farhoomand, A., and Ho, P., "Innovation through Supply Chain Reconfiguration," *MISQ Executive* (3:3), 2004, pp. 131-142; Loebbecke, C., "Modernizing Retailing Worldwide at the Point of Sale," *MISQ Executive* (3:4), 2004, pp. 177-187; Rockart, J., "Information: Let's Get It Right," *MISQ Executive* (3:3), 2004, pp. 143-150.

[4] Jasperson, J., Carter, P. E., and Zmud, R. W., "Post-Adoptive Behaviors Associated with IT-enabled Work Systems," *MIS Quarterly* (29:2), 2005, pp. 525-557.
[5] Data warehousing refers to the family of technologies associated with creating repositories of cleansed, formatted, and well-integrated data, and enabling business professionals to access and analyze that data to produce routine and ad hoc reports.
[6] Business intelligence involves collecting data, analyzing the data to detect patterns and meanings within the data, extracting information from these analyses, and turning this information into actionable knowledge. With advanced technologies (data warehouses, enterprise systems, sophisticated analytical tools, etc.), the volume and breadth of data available for analysis is growing rapidly in most organizations.

tions can take a variety of forms, three environments are most common:[7]

1. *A shared service* made available to users across an enterprise. This consists of a central BI capability that includes a single data warehouse, analytical tools, and a professional BI staff skilled in accessing the data and using the tools. Generally, users work through this BI staff to satisfy their BI needs.

2. *Targeted data marts*, which are focused data warehouses built to meet the BI needs of a specialized user group (e.g., market research analysts, supply chain professionals, etc.). After receiving appropriate training, these user groups are typically able to satisfy their BI needs themselves using their data mart.

3. *Menu-driven reporting systems*, which enable the firm's business professionals to access a data warehouse to produce standard reports and to initiate simple queries. When such reporting systems are targeted at a firm's executives, they are typically referred to as executive information systems.

However, we have observed a broader approach to BI, which, though still quite rare, could well become its next phase. In this approach, an enhanced data infrastructure is combined with an information culture so that business professionals can regularly access the data warehouse, in collaboration with others, to create business solutions to relatively unique problems and opportunities. We call this approach "advanced business intelligence." We view advanced BI as involving two key components:

1. *A data infrastructure*, which includes the data warehouse, an enterprise-wide data model, data access and analysis tools, and a support infrastructure. Together, these components provide universal access to a common set of data by all business professionals.

2. *An information culture*, where business professionals are expected to justify their decisions and plans through "hard" data and to proactively meet their own BI needs.

This article examines Cardinal Health's efforts in the 1990s to evolve such an advanced BI environment, it describes how that environment is being used, and it offers recommendations on how to build such a capability.

CARDINAL HEALTH'S MEDICAL PRODUCTS AND SERVICES BUSINESS

Cardinal Health is a global provider of integrated solutions for the healthcare industry. The company's businesses distribute and manufacture pharmaceutical and medical products, and offer a range of services that improve customers' clinical and financial performance. The subject of this study is Cardinal Health's medical products and services business, which includes the largest U.S. manufacturer and distributor of medical, surgical, and laboratory products and services used by healthcare providers. This group of businesses manufactures products such as surgical instruments, drapes, gowns, and gloves. And, its supply-chain arm distributes medical, surgical, and laboratory supplies from over 2,600 manufacturers to nearly 80,000 sites of care.

This segment of Cardinal Health had a rich and successful heritage, beginning with the 1922 founding of American Hospital Supply Corporation (AHSC). In the 1980s, AHSC gained a reputation as an early IT innovator and one of the few companies to gain competitive advantage from its use of IT. Its ASAP system revolutionized supplies distribution to hospitals.[8] In 1985, AHSC was acquired by competitor Baxter International, and then spun off by Baxter in 1996 as Allegiance Corporation. Allegiance was acquired by Cardinal Health in 1999 and formed the core of Cardinal Health's medical products and services business.

This heritage has three direct influences on the organizational environment of the medical products and services business. First, most executives and senior managers, as well as many employees, have long tenures. Thus, its business strategy is sales and marketing oriented and its decision-making is financially and analytically oriented, that is, data driven. Second, its culture is highly receptive to IT-enabled business innovation. Third, due to the long employee tenures, personal networks are dense and rich. People across the firm tend to trust each other because they either know each other or know *about* each other.

Despite this lengthy heritage, there is little evidence of an entrenched bureaucracy. Allegiance was essentially "reborn" when it was spun off from Baxter International in 1996. At that time, the healthcare industry was undergoing dramatic market changes: powerful group purchasing organizations emerged and margins narrowed. As a result, sales compensation structures changed. After the spin off, Allegiance Corporation

---

[7] Gardner, S. R., "Building the Data Warehouse," *Communications of the ACM* (41:9), September 1998, pp. 52-60.

[8] Harvard Business School Case #9-186-005, "American Hospital Supply Corp. (A): The ASAP System," HBS Case Services, Boston, Mass., 1985 (revised 4/1986).

became a very lean organization to ensure the company's survival. Thus, the stage was set for rather dramatic organizational change initiatives that, in more normal times, might have been resisted.

In addition, Allegiance's IT decision-making processes were noteworthy, as were the IT organization's role and structure. IT decisions were largely driven by senior executives and line managers, all of whom recognized the important role of IT and the need to continue to invest in and innovate with IT. The IT organization's role, thus, was to facilitate and support line management in making these IT-related decisions. Information, IT assets, and IT professionals were seen as key business resources that needed to be applied regularly and well. IT operating expenses were charged back to operating units, largely on an overhead basis, and accepted as a normal cost of business. IT capital investment was treated like any other capital investment – it was financially driven. However, the IT organization was provided with investment funds to enrich the IT infrastructure. (These investments were also justified, in the same manner as with any other capital investment.) Through careful management of these funds, monies were available for "seeding" IT experiments and prototyping systems.

Most of the IT professionals also have long tenures with the company, especially those in senior positions. As a result, the IT organization as a whole has a deep understanding of the business and its business processes. The IT professionals are located both at an enterprise level (focused on enterprise-wide IT activities) and within business units. In fact, most business functions are technologically supported by both co-located IT professionals and an IT-savvy set of business analysts. These business analysts serve as the primary day-to-day internal IT consultants to a unit's business professionals. The co-located IT professionals, on the other hand, are mainly charged with handling a unit's major technology initiatives and resolving technology-intensive operational and use issues.

THE SAP R/3 AND DATA WAREHOUSE INITIATIVES

In 1995 (just prior to the Baxter spin-off), Baxter senior management decided to implement SAP R/3® to accomplish three objectives: resolve Y2K issues, replace outdated legacy systems (multiple databases with similar but unsynchronized data, inferior user interfaces, unconnected systems, etc.), and decommission some 20 end user computing systems (which were effective within single business units but created huge information challenges across units). Planning for the SAP implementation and its associated business process reengineering efforts occurred during 1995 and 1996. The $64 million SAP implementation took place in 1997 and 1998. Figure 1 shows the goals of this initiative, which provided the foundation for Cardinal Health's current advanced BI capabilities.

Figure 1: SAP / Data Warehouse Project Goals

- Create a single source for all data.

- Achieve a high responsiveness with business transactions.

- Improve quality & currentness of data.

- Switch reporting mindset from push to pull.

- Reduce business reliance on IT.

- Lower IT operations (hardware, software, people) costs.

Using the Data Warehouse as a "Reporting System"

During the planning, management made the decision to use a data warehouse, rather than the SAP reporting system, for all reporting. At that time, SAP did not have a data warehouse product. Management based its decision on three main factors:

- The many stories in the business press about ERP failures gave management a strong incentive to simplify the SAP implementation as much as possible.

- Management was concerned that SAP could not handle the expected high transaction processing volumes. At the time, this installation was the largest single instance of SAP. Separating transaction processing and business reporting would improve transaction processing responsiveness and would also protect the operational systems (which change infrequently but are expensive to change) from the reporting systems (which change frequently and are less expensive to change).

- Finance management was not satisfied with the report generation tool in SAP at that time. They believed it was inferior to the McCormick & Dodge report writer that the finance function was then using.

Thus, management decided to build a data warehouse using purchased software components where possible.

Figure 2 provides an overview of this solution. As shown, it included three client interfaces to serve people in three very different roles. The first was a hierarchically-structured, menu-driven tool for the sales force. The second was a Web-based front-end for customers and suppliers. Both of these interfaces were quite restrictive, but very intuitive and, hence, easy to use. These two interfaces were targeted at client segments where on-going investments in training and support would probably not pay off. The third interface – built around the Business Objects® tool set – was for all other employees. Once trained, employees could use Business Objects to make either simple or sophisticated queries of the data warehouse. Business Objects was powerful enough to support both.

Figure 2: Data Warehouse Design

Component	Description
Database	• Oracle
Interface	• Homegrown
Client Tool Set	• SAP single table inquiry tool • Business Objects tool set • Download capability into Microsoft Access, SAS, and Excel
Client Interface	• Hierarchically structured (for sales force) • Web-based front-ends (for customers and suppliers) • Business Objects tool set (for most employees)
Data Universes	Sales history, invoice, inventory, purchase order, rebate, pricing, accounts receivable, accounts payable, financial, fixed assets, human resources

Reengineering the Reporting System

As part of the overall business process reengineering effort that accompanied the SAP implementation, Cardinal Health formed teams of business analysts, users, and IT staff to create a uniform, simplified reporting structure across the firm. The IT members of these teams had previously been involved in end user computing, so they were adept at working with end users and understood their reporting needs. These teams spent most of their time first listening to business users, and then designing a minimal set of master report templates, which the company called "static"

templates. Largely by saying, "No!" these teams produced a simple reporting structure that met most individuals' reporting needs – and eliminated thousands of former reports.

The results were (1) a set of stable, repeatedly used Business Objects templates (the static templates), and (2) an ad hoc reporting capability, whereby a user could modify an existing Business Objects template or build a new one to access data in the warehouse. The business units, not the IT organization, "owned" the master static templates. The "new" templates were stored locally and retrieved with refreshed contents. However, individual users could not modify the master static templates without following the company's formal modification process.

The new static and ad-hoc reporting structure changed Cardinal Health's reporting from a "push" system to a "pull" system. The IT organization no longer needed to create and send reports to individuals. Instead, employees accessed the information they needed, either by executing an existing template, by modifying and then executing the modified template, or by creating and then executing a new template.

Data Warehouse Marketing and Training

Up to 50% of the data warehouse implementation effort was spent communicating with employees about the data warehouse, and then training them how to use it. Thousands of employees attended a 2½ day basic class (to obtain their Business Objects ID). They could also attend annual update classes and more advanced classes on specific areas. A five-person Business Objects support group provided the training, handled general support, and maintained a Web site that communicates "news" about the data warehouse.

Creating Finance Super Users

The finance function originally housed all the data warehouse "finance super users." These users were financial analysts with three to four years of experience, who were chosen by the directors of finance to receive intensive training on the warehouse. They became the data warehouse experts, and made the most sophisticated use of it. The role "financial super user," was, in fact, formally defined, but it was not a formal job assignment. These analysts held regular assignments as well. But they met regularly, their supervisor gave them the time to perform their super user role, and their annual evaluation included their performance in this role.

Due to the importance of financial data across Cardinal Health, each "finance location" (headquarters, division, and office) was required to have both a finance super user and a backup super user. Thus, finance "seeded" the entire enterprise with "finance super users." The initial result was some 60 finance super users throughout the company, each of whom could exploit the data warehouse for their local business unit by carrying out sophisticated analyses themselves, and by helping co-workers understand the data, access and execute static reporting templates, and modify and create ad hoc reporting templates. These super users are viewed by other employees as their unit's data warehouse experts.

Over time, the composition of these sophisticated data warehouse users has changed. While the first ones were chosen by supervisors, as they moved on to new assignments, they selected and mentored their replacements. The result has been an ever-enlarging community of highly knowledgeable data warehouse users.

Turning Ownership of Data and Report Templates Over to the Business Units

The business units historically "owned" the data in the data warehouse, but not maintenance of the data or the reporting systems. Why? Because the IT group, rather than the business units, had the expertise to modify the data and the reporting structures. However, that has now changed due to the dispersion of the data warehouse expertise. IT is still responsible for populating the data warehouse cube structure, but the business units now have the abilities and tools to handle both data and template maintenance. So, they now "own" both of these maintenance functions.

Turning Off the End User Computing Systems

As data moved into the data warehouse, the end user computing systems were depopulated, and the data and reports became accessible only through the data warehouse. This change forced the business users to embrace the data warehouse – it became the only game in town. As a consequence, the newly developed data and templates enforced the "pull" reporting structure, and also became the common business language.

Evolving the Data Warehouse

From the start, business and IT professionals viewed the data warehouse as an evolving entity, not a one-time event. They realized it would be ever-changing to continually increase its functionality and scalability. As a consequence, the processes for improving and extending the data warehouse have become an accepted part of organizational life, and employees across the enterprise participate in them.

With these capabilities in hand, and widespread, Cardinal Health has moved beyond traditional uses of data warehouses and into advanced business intelligence.

FIVE ADVANCED BUSINESS INTELLIGENCE PROJECTS AT CARDINAL HEALTH

Cardinal Health's advanced BI capabilities are exercised through the on-the-fly formation of teams of people who have the expertise and capabilities to resolve a problem or exploit an opportunity. Generally, these teams do not use the existing report templates, but rather extend BI by inventing new ones to fit their specific needs. The general three-step BI process is as follows:

Step 1: Assess whether or not the data warehouse might be useful in producing a solution to a problem or an opportunity,

Step 2: Identify the expertise and capabilities needed to produce the solution, and then

Step 3: Pull together a team of individuals from the appropriate functional areas, perhaps including a financial super user or an IT professional.

This ad hoc team structure has made advanced BI possible. Generally, only one team member works with the data warehouse, using Business Objects to perform data retrieval and analysis. Often, the extracted data undergoes additional analysis using Excel®, MS-Access®, or SAS®.

Following are five examples of advanced BI work at Cardinal Health. They illustrate the wide variety of work in terms of business functions, problems, and opportunities. They range from a large, enterprise-wide customer support system to a single special-purpose report (see Figure 3).

Creating a List of Inventory Items to Drive Sales for Care Continuum Customers

During a regular meeting of the care continuum management team in the Southwest region, the members, once again, noted the growing problem of excess inventory. (Care continuum customers represent one of the firm's market segments.) Initial analysis indicated that the problem was caused, in part, by stocking items with multiple substitutes. The proposed solution was to create a list of items that met the majority of these customers' requirements, then guarantee availability of those items, and have the sales force focus on them when interacting with customers. But, what

Figure 3: Five Advanced BI Projects	
Project	**Description**
Product List	Identified "best value" inventory items guaranteed for availability.
Delivery Plant	Located alternate warehouses for locally out-of-stock items.
Entelligence	Created a Web-based purchase management system that allowed customers to take advantage of data warehouse capabilities.
Standard Reports	Consolidated and standardized HR reporting templates for company-wide use.
Supplier Diversity	Identified suppliers who met the Federal Government's small business diversity requirements.

items should be on this list? The region's inventory coordinator believed the data warehouse could help solve the problem, so a team was formed with individuals from operations, inventory, and accounts management.

By examining the products flowing from the region's plants, the team created an initial list of items. Nine sales people then tested the list with clients, and found that it was not broad enough because it did not reflect items shipped from outside the region. Further analysis generated a second list, which was again tested by some sales people. In the end, the list was further refined by listing only "best value" products, that is, products that represented best value for the money.

The care continuum sales team then used this list to drive sales. Ordering from the list provided customers with better fill rates because all the items were guaranteed to be in-stock. Later, redundant items were deleted.

The region's operations manager made the list available to her peers at a regular meeting. The concept and the process became a model for the enterprise. Later, all but 400 of the 1,300 items on the region's list appeared on the national list.

Identifying an Alternate Delivery Plant to Ensure Fast Product Delivery

An inventory manager realized that orders from one customer were failing because items were out-of-stock at that customer's "normal, local" delivery plant. She envisioned a solution of rerouting these orders to an alternative delivery plant that had the stock in inventory. By speaking to customer service representatives and materials management experts, she verified that automatically re-routing these orders would not create operational problems.

She knew that the data warehouse contained the data to solve the problem. So she enlisted the help of the materials management IT group to build a solution. Initially, they created a test file from SAP R/3 for her to manually determine the default alternate delivery plant. Then, the IT materials management group modified SAP R/3 to automatically use this alternate shipping location when the normal location was out-of-stock.

Her solution reduced the manual work associated with these orders. In addition, Cardinal Health lost fewer sales, had fewer drop-shipments and returns, and even reduced inventory levels. The inventory manager shared her solution with other divisions, which applied similar logic to their business. In fact, IT and materials management initiated a formal project to apply this logic to filling orders for items that are active in some regions but discontinued in others.

Giving Customers Access to Their Own Data through the Entelligence Product

With the rise of Internet-based commerce, the e-business team at Cardinal Health wanted to give customers Web-based access to their own data in the Cardinal Health data warehouse. The goal was to foster greater customer satisfaction and retention by providing them with a tool for investigating ways to reduce their costs by modifying the nature, timing, and volume of their purchases.

A team with members from e-Business and IT was put together, based on their knowledge of the data warehouse and the customers. They first created a logical data warehouse for each customer, which excluded sensitive data, simplified the reporting structure, and restricted customers to their own data. The team then built a prototype and pilot-tested it by bringing customers on campus. It then solicited pilot sites at customer organizations. The number of pilot sites grew from 4 to 25 prior to full roll-out.

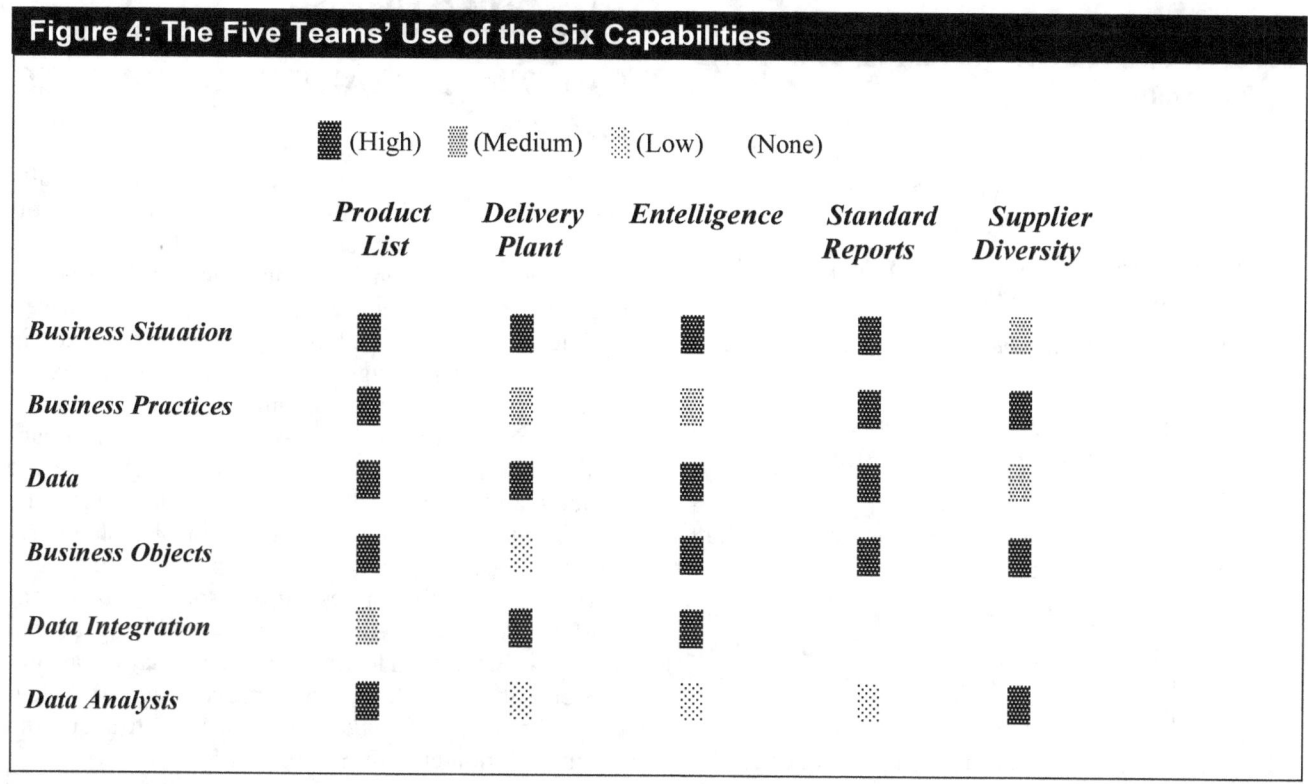

Figure 4: The Five Teams' Use of the Six Capabilities

Cardinal Health called this new product "the Entelligence Purchase Management System." In its initial 18 months of operation, the user base grew to 600 customer organizations. The product generated subscription fees, reduced customer churn, and provided cost savings by reducing the number of work-hours needed to support customers. Entelligence continued to evolve as the group's e-business vision evolved and as the regional e-business managers and Cardinal Health's overall e-business team reviewed customers' and the sales force's requests for changes.

Complying with a Corporate Mandate through Standard Report Templates

When Cardinal Health acquired Allegiance Corporation in 1999, it required all Allegiance divisions to regularly provide a wide variety of human resource (HR) information. Because Allegiance was not using SAP R/3's report generator, each division needed a report template to comply with this corporate mandate. A team of HR business analysts and members of the co-located HR IS unit undertook the task because they were involved in the data warehouse development effort, they understood the data, and they knew how to access the data using Business Objects templates.

By surveying the HR staff in each division, the team determined the data each staff wanted to "pull" from the data warehouse. An HR business analyst then produced master templates, which division HR staff could pull from the "corporate documents" area of Allegiance's intranet. The team then trained the HR employees to access and modify the templates to suit their division.

Reporting on Supplier Diversity for the Federal Government

To comply with government contracts, Cardinal Health must file supplier-diversity reports quickly and accurately. These reports provide statistics on the company's small or minority-owned business suppliers. They draw data from numerous areas of the data warehouse, including archived balance sheet and accounts payable data. To create these reports, the company established "rules" to accurately identify all suppliers who qualified and were certified as small or minority-owned businesses. The reports were produced by a team of financial analysts from corporate finance because they knew the data in the data warehouse, as well as how to get data from the various business units.

Once the team finalized the template for these reports, which included over 180 queries, it was produced and accepted by the regulatory body. Then, financial analysts in the small business unit were trained how to use the template and were made responsible for modifying it in the future.

THE NATURE OF THE ADVANCED BI TEAMS

These five teams were clearly quite different from one another. However, they did possess six common capabilities:

1. comprehension of the *business situation* (problem or opportunity) to be addressed,

2. understanding of the *business practices* associated with the situation,

3. ability to locate and retrieve needed *data* (via the data warehouse or social networks) in order to resolve the situation,

4. proficiency with *Business Objects*, to access data in the data warehouse,

5. expertise in *data integration*, to pull together and work with data from a variety of sources (data warehouse, SAP R/3, other information systems, etc.), and

6. adeptness in *data analysis* (using Excel, MS-Access, or SAS).

Figure 4 shows each team's dependence on each of these capabilities.

As shown, capability use varied by project. For the most part, all the capabilities were provided by the business professionals on the team, not by IT professionals or data warehouse specialists. While teams were needed to garner all the needed skills, it appeared that each team's performance was greatly enhanced by including a *single* individual with both business practice and data access capabilities, or with both data availability and data access capabilities.

We were especially impressed with three attributes of Cardinal Health's BI teaming process. The first was each project leader's ability to identify individuals across the enterprise who had the knowledge to solve the problem at hand – and then get them onto the team. Second, we were impressed with the team members' ability to identify, access, and acquire the needed knowledge, no matter where it existed in the enterprise. Third, we were impressed with each team's ability to diffuse its learnings to others. Importantly, Cardinal Health had no formal mechanisms to facilitate these behaviors. Instead, they seemed to occur naturally, as a consequence of Cardinal Health's organizational infrastructure.

THE NATURE OF THE ORGANIZATIONAL INFRASTRUCTURE ENABLING CARDINAL HEALTH'S ADVANCED BI CAPABILITY

An "organizational infrastructure" is conceptually similar to a "technical infrastructure." Just as the effectiveness of an application system depends on the robustness of its enabling technical infrastructure (e.g., desktop clients, servers, networks, databases, middleware, etc.), the effectiveness of employees in carrying out their assigned duties depends on the robustness of their enabling organizational infrastructure (e.g., values, norms, rules, resources, etc.). An organizational infrastructure develops as a function of the policies and practices that frame organizational life, both formal (evaluation systems, control systems, resource allocation procedures, HR practices, etc.) and informal (leadership styles, supervisory feedback, coworker interactions, etc.). Our analyses identified two aspects of Cardinal Health's organizational infrastructure core that are relevant to employees' ability to leverage the data warehouse: the data infrastructure and the information culture. See Figure 5.

Figure 5: Organizational Infrastructure Enabling Cardinal Health's Advanced BI Capability

- Data Infrastructure
  - an enterprise-wide data model
  - an enterprise-wide, limited set of data access tools
  - a robust support environment
- Information Culture
  - a data-driven decision style
  - business-led IT decision-making
  - dense and robust social networks
  - "pull" reporting structures

The BI Aspects of the Data Infrastructure

Three aspects of Cardinal Health's data infrastructure have most contributed to its advanced BI capabilities. The first has been its *common data model* – housed in the data warehouse. This model established a common business language across the medical products and services business. Because of this data-driven common business language, Cardinal Health's business professionals now share a common understanding of the firm's business metrics, so that in their daily work, they can see the business performance of functional units and the overall enterprise. With these metrics and the common language, employees can more easily collaborate, as exemplified by the advanced BI projects.

The second contributing aspect of the data infrastructure has been the decision to *limit the variety of end user tools*. Everyone is trained on the same tool set. Even though they are at different skill levels, everyone speaks the same "data warehouse language." Furthermore, by limiting the number of tools, Cardinal Health has optimized its spending on training and support.

The third contributing aspect of the data infrastructure has been the *robust support environment*. Early on, management recognized the importance of sufficiently funding communication, training, and creation of a distributed support structure. The communication and training created a huge base of expertise across the enterprise, all of which can be brought to bear when problems or opportunities arise. The distributed support structure has proven equally valuable because all Cardinal Health's business professionals know they can get sound support locally and quickly if they encounter difficulties or complexities "pulling" or manipulating data. They know that co-workers, super users, their unit's business analysts, their unit's IT support group, and even corporate IT are all there to help. This fail-safe support environment has prompted Cardinal Health's business professionals to not hesitate in tackling innovative, complex, and advanced BI endeavors.

The BI Aspects of Cardinal Health's Information Culture

Cardinal Health's information culture has four important advanced BI aspects: a data-driven decision style, business-led IT innovation, dense social networks, and a pull reporting structure. The first three reflect values and norms stretching over three decades. The last component – pull reporting – was developed more recently, as a result of its decision to eliminate all "push" reporting systems during SAP R/3 implementation.

The first BI-relevant aspect of its culture is that all key *decisions are data-driven* to the extent possible. This means that decisions are based on solid and confirmable marketing and financial data. As a consequence, accessing and applying stored data to support business decisions and actions are second nature for the business professionals.

The second important cultural aspect is that *IT decisions are business led*. The vast majority of decisions about how, when, and where to use IT to enable or support business initiatives and processes are initiated and led by business professionals. However, this does not mean that the company's IT professionals are passive agents. Rather, they are actively involved in innovating with the business professionals. The business professionals view the firm's IT professionals as highly knowledgeable about the business, as highly capable of delivering on their promises, and as exceptional custodians of IT resources. IT staff are seen to provide increased service levels while keeping the IT budget flat or below the inflation rate. What is perhaps most telling is that the question of "IT's value to the business" seldom arises. Instead, the value of IT is implicitly addressed during normal business decision-making. Senior executives understand the need to invest in and innovate with IT.

The third cultural contribution to Cardinal Health's advanced BI capabilities is its *dense social networks*. Most employees have deep and wide social networks because the company has had such low turnover and because it has a bias for moving employees across functional and geographical boundaries to progress their careers. As a result, employees know many other employees, and what those employees do and know. Hence, they can quickly and accurately identify who might possess the expertise they need for an advanced BI project.

Finally, the fourth important contributor to Cardinal Health's advanced BI culture is its *"pull" reporting structure*. Data can only be obtained by proactively pulling it from the data warehouse, either by using an existing Business Objects template, revising an existing template, or creating a new template. Thus, using the data warehouse is part of how employees work. It's a routine aspect of the company's operations. Essentially, employees are personally responsible for crafting their own information environment in carrying out their work.

RECOMMENDATIONS

One advantage of data warehouses is that they can be designed to not constrain data queries and retrievals to designers' original intentions. As a result, data can be extracted, summarized, and combined by users to discover solutions to problems the design team did not envision. As the Cardinal Health case shows, a data warehouse can create the foundation for advanced BI work. However, the enterprise must also commit to the warehouse being an enterprise-wide data resource with tools for accessing it, and to informing and training employees on this resource. Many organizations have, in fact, created such a data warehouse capability. However, few have evolved it into an enterprise-wide advanced BI capability, as Cardinal Health has.

But, as the Cardinal Health evolution has shown, business intelligence does not reside in the data warehouse. But rather, BI emerges in the minds of employees when they identify and access data, combine data with their own or others' knowledge of a business situation, and produce novel resolutions to the issue at hand. In our view then, a company's organizational infrastructure – its data infrastructure and information culture – is the core driver of advanced BI capabilities. With these structural components in mind, we offer the following three recommendations for developing advanced BI capabilities.

Recommendation #1: Establish an organizational climate where business professionals are expected to be in command of their own information destiny.

"Being in command of one's own information destiny" means that all employees understand that it is their personal responsibility to actively participate in building an information environment to support their own decision-making and problem-solving. Cardinal Health's business professionals cannot expect reports to periodically be provided to them. Instead, they must retrieve (either by themselves or working through a co-worker) the desired information by executing an existing template, modifying and then executing an existing template, or creating and then executing a new template.

To establish such an organizational climate, two elements must come together. First, employees must see value in basing their decisions on hard data. If an organization's decision style is not fact-based, then employees will likely not make the effort to understand, find, or access the data and metrics important in their work. Second, employees at all levels need to be technology-savvy. While many technology issues should be left to the experts, when an organization's employees are technology-complacent, they are not aware of available technologies. More to the point, they are not technologically curious, so they fail to ask, "I wonder if this is technologically possible?" Having business professionals willing to "push the envelope" technologically is a key attribute of the advanced BI teams at Cardinal Health.

Over time, Cardinal Health's organizational culture has pushed its business professionals to take control of their own information destiny. They are expected to support decisions using hard data, and actively participate in building their own personal information environment. While such a culture cannot be built quickly, organizations can take the following three actions to move in this direction:

1. Decentralize IT and data governance structures because decentralization broadens awareness, familiarity, and competence with IT functionality and data availability.

2. Build up a rich portfolio of metrics to assess and direct business decisions, actions, and behaviors.

3. Reduce the number of automatically distributed standard reports while still requiring employees to perform at high levels.

These three actions will expand the use of an existing data warehouse by encouraging greater use in everyday work.

Recommendation #2: Encourage rich social networking across the organization.

An advanced BI capability depends on employees developing deep knowledge of where expertise exists in their enterprise. Most business professionals do not have personal access to all the data and knowledge needed to effectively resolve all the problems and opportunities they confront. Therefore, they need to identify, contact, and collaborate with others to get that data and knowledge. Generally, to find expertise, people ask someone they know. If they know many people, they have a dense social network. If they also know about what those people know, they have a robust social network. Without dense, robust social networks, we believe advanced BI efforts will bear little fruit because such efforts depend so heavily on quickly and collaboratively assembling the needed expertise.

Most Cardinal Health employees are well-positioned in numerous social networks that have been nurtured over many years. While the drivers of these social networks are rather unique (long-tenured employees who have lived through cross-functional and cross-unit career paths), there are nine actions enterprises can take to foster social networking:

1. Consider the long-term impacts of planned workforce reductions – even though they clearly have very attractive short-term impacts in tough economic times.

2. Value tenure during workforce reductions; in many companies, the most experienced and well-connected employees are often the first to be let go.

3. Orchestrate cross-functional and cross-unit career paths so that, over time, employees are able to work with many different people.

4. Make extensive use of cross-functional and cross-unit teams.

5. Use virtual teams rather than co-located teams.

6. Commingle workspaces.

7. Implement knowledge management capabilities.

8. Implement enhanced communication and collaboration capabilities.

9. Hold formal and informal meetings to bring together diverse employee communities.

Recommendation #3: Establish a distributed support infrastructure that emphasizes local "first-responders."

Advanced BI requires, at the least, vigorous use of a data warehouse. A vibrant advanced BI capability will not materialize where business professionals use their data warehouse timidly – rather than stretching it and molding it to their needs. Unless they know that help is always nearby, employees are unlikely to aggressively use a data warehouse by creating new, unproven accesses.

At Cardinal Health, employees just expect super users and unit support staff (and, if needed, unit and corporate IT professionals) to be available to answer their questions or provide expertise. This depth of data warehousing expertise has provided a fail-safe environment, which means that there is always someone who can get business professionals out of any data warehousing situations they get themselves into. With this assurance, the business professionals are not timid in their uses of the data warehouse.

Cardinal Health's support environment is unique because it is largely the responsibility of the business units, not IT. Corporate IT nurtures and supports the environment, but the business professionals who serve as the experts and super users are the first-responders. While such a support infrastructure cannot be quickly established, organizations can take four actions to develop it:

1. Identify business roles that both are broadly distributed across the organization and would benefit from intensive data warehouse use. Leverage these roles within support infrastructures.

2. Provide universal data warehouse education and training.

3. Provide a highly visible, accessible, and informal problem escalation process, such as from super user to unit business analyst staff to unit IT staff to corporate IT.

4. Include in performance evaluations an individual's responsiveness to others' formal or informal requests for data warehousing help.

CONCLUSION

In today's world of global hyper-competition, companies want to see demonstrable results from their use of information technology. Many organizations have invested in data warehousing to improve their competitive position. One view of a data warehouse is as a "queryable source of data in the enterprise."[9] That is, it is a product delivered by information technology to the organization. Another view is to see the data warehouse as a component of an advanced business intelligence capability, as we have described in this article. As such, the data warehouse becomes integral to the competitive capacity of the business.

One of the unique features of Cardinal Health's data warehouse and BI capability is that no one thinks of it as such. Instead, employees see it as one of many tools for solving business problems. As the integrity and timeliness of the data have improved, and as employees have been able to easily access and integrate financial, sales, and logistics data, they are creating the metrics and the business intelligence they need to build innovative solutions to the business situations they face.

ABOUT THE AUTHORS

Traci A. Carte

Traci A. Carte (tcarte@ou.edu) is an Associate Professor of MIS in the Michael F. Price College of Business at the University of Oklahoma. She received her Ph.D. in MIS from the University of Georgia. Currently, her research interests include IT support for diverse teams, politics and IT, and research methods. Her research has been published in such journals as *MIS Quarterly, Information Systems Research, Decision Support Systems,* and *Journal of the AIS.* She serves on the editorial board of *MIS Quarterly.*

Albert B. Schwarzkopf

Albert B. Schwarzkopf (aschwarz@ou.edu) is Associate Professor, MIS Division, Michael F. Price College

---

[9] Kimball, R., Reeves, L., Ross, M., and Thornthwaite, W., *The Data Warehouse Lifecycle Toolkit: Expert Methods for Designing, Developing, and Deploying Data Warehouses,* John Wiley and Sons, New York, 1998, p. 19.

of Business, University of Oklahoma. His research interests focus on end user access and use of data and information. He holds a Ph.D. in mathematics from the University of Virginia.

Teresa M. Shaft

Teresa M. Shaft (tshaft@ou.edu) is an Associate Professor of MIS at the University of Oklahoma's Michael F. Price College of Business. She received her Ph.D. in Management Information Systems from the Pennsylvania State University. Her research interests include the cognitive processes of systems developers, IT effectiveness, and the role of information systems in environmental management. Her research appears in journals including *Information Systems Research, Management Information Systems Quarterly, Journal of Management Information Systems, and Journal of Industrial Ecology.* She was a co-founder of the AIS Special Interest Group IS-CORE. Her research has been supported through grants from the U.S. National Science Foundation.

Robert W. Zmud

Robert W. Zmud (rzmud@ou.edu) is Professor and Michael Price Chair in MIS, MIS Division, Michael F. Price College of Business, University of Oklahoma. His research interests focus on the organizational impacts of information technology, and the management, implementation, and diffusion of information technology. He currently is a Senior Editor with *Information Systems Research* and *MISQ Executive.* He also sits on the editorial boards of *Management Science, Academy Management Review,* and *Information & Organization.* He previously held the positions of Editor-in-Chief of *MIS Quarterly* and Research Director of the Society of Information Management's Advanced Practices Council. He is a fellow of both AIS and DSI. He holds a Ph.D. from the University of Arizona and an M.S. degree from MIT.

APPENDIX: RESEARCH APPROACH

Our understanding of the Cardinal Health approach to business intelligence was derived from over 60 interviews conducted by the research team both on-site and via telephone. Our investigation consisted of four phases. In Phase I, we interviewed the IT leadership team to understand the firm, its business environment, and the motivation and process for its BI efforts. In Phase II, we interviewed senior business managers across the firm to understand the business value they have derived from the BI capability, as well as how they were exercising this capability daily. In Phase III, we conducted telephone interviews with approximately 35 mid-level business managers across the firm to identify examples of BI initiatives. From that sample, we selected five for more intensive examination. In Phase IV, we conducted face-to-face interviews with the members of these five teams.

PROFIT FROM CUSTOMER DATA BY IDENTIFYING STRATEGIC OPPORTUNITIES AND ADOPTING THE "BORN DIGITAL" APPROACH[1]

Gabriele Piccoli[2]
University of Sassari

Richard T. Watson
University of Georgia

Executive Summary

We present a framework that maps the four data-driven strategies—Minimize Costs, Reward Loyalty, Personalize Interactions, and Acquire Customers—that a firm can enact to extract value from its customer data. The four strategies are distinguished by the potential repurchase frequency and the customizability of a firm's products or services. We describe each of the strategies and provide in-depth examples from companies in the hospitality industry. By positioning themselves within the four-strategy framework, firms in a wide range of industries will be able to envision how they can adopt the most appropriate strategy (or strategies) for exploiting customer data to improve profitability.

We also discuss the importance of "born digital" data, whereby data is captured in digital form, not digitized through scanning or manually input. A proactive born digital approach enables firms to better exploit opportunities for extracting business value from customer data.

THE NEED FOR DATA-DRIVEN CUSTOMER STRATEGIES

Whether a profit-oriented enterprise focuses more on recruiting new customers, retaining current customers, or both, is significantly determined by the *potential repurchase frequency* of its products or services. If there are few returning customers for a product or service that has a potentially high repurchase frequency, it is a clear signal that the company is missing an opportunity. However, firms with products or services that have potentially low repurchase frequencies need to focus their attention on acquiring new customers.

In addition, some products and services can be tailored to the specific needs and preferences of individual customers or customer segments. The *degree of customizability* of a product or service is related to its complexity. For example, Canyon Ranch (a U.S.-based health resort and hotel chain) offers over 230 spa, health, and healing services that now include medical laboratory tests worthy of some of the most advanced hospitals in the country. Because of the wide array of options and the ability to tailor services to the unique needs of a guest, every Canyon Ranch guest "experiences a different Canyon Ranch," as the firm is fond of saying.[3]

Based on these two dimensions—potential repurchase frequency and product or service customizability—we propose a framework that identifies four strategies for exploiting customer data. We describe each strategy in turn, illustrating them with examples from companies in various segments of the hospitality industry, based on

MISQE is
Sponsored by

KELLEY SCHOOL
OF BUSINESS
INDIANA UNIVERSITY

Delivering Business Value
Through IT Leadership

---

1 Carol Brown is the accepting Senior Editor for this article.
2 The authors would like to thank David Sjolander and Paolo Torchio for their valuable comments on earlier versions of this article.
3 Applegate, L. M., and Piccoli, G. *Canyon Ranch*, Case #805027, Harvard Business School Publishing, 2004.

our in-depth research in that industry.[4] The hospitality industry provides a good test bed for our ideas because the service experience leads to the generation of significant customer data. The ideas advanced in this article, however, are not unique to the hospitality industry.

The strategic opportunities for profiting from customer data obviously require a firm to capture high-quality data about its customers. The Internet and the pervasive computerization of business activities enable most customer data today to be "born digital"—i.e., captured directly in digital form at its inception and not manually input in computer systems or digitized through scanning a document. The adoption of customer relationship management (CRM) and business intelligence (BI) applications has often focused on using customer data collected as a byproduct of transaction processing systems. We believe companies should take a more proactive approach and consciously plan to capture data in digital form to enable more effective data-driven customer strategies.

We conclude this article by identifying the lessons that can be learned from the hospitality industry examples and presenting guidelines for applying the four-strategy framework to develop proactive born digital data strategies in other industries.

FOUR STRATEGIES FOR EXPLOITING CUSTOMER DATA

Our framework is predicated on recognizing that the business value of customer data can be assessed from the characteristics of the goods or services that every firm specializes in providing. Barring change of product mix or significant innovation, these characteristics are fixed and can be categorized on two principal dimensions: potential repurchase frequency and degree of customizability. As shown in Figure 1,

categorizing products and services in this way gives rise to four strategies for exploiting customer data.[5]

1. Minimize Costs

A firm should focus on minimizing costs when there is little likelihood of repeat business and few options for customization. Transactions in such firms should be efficient and uneventful. The customer is served at minimum cost and neither party expects to have many future encounters.

In this type of business, there seems to be little potential strategic opportunity for exploiting customer data, at least when focusing on an individual product or service rather than the opportunities for bundling. There is little that managers can do with the customer data to increase loyalty or customize the product. Consider, for example, a chain of budget or limited-service tourist hotels in an exclusive fly-in destination, such as Hawaii or Fiji. Mid-scale hotels in these locations are generally a "window on an experience" rather than the experience itself, and the chain's value proposition is offering guests an affordable opportunity to experience a great location. Because of the time commitment and cost of reaching these destinations, repurchase is relatively infrequent. Under these conditions, the chain is better off focusing on efficiency and low prices. Figure 2 describes how one such chain—OHANA Hotels—deliberately decided not to adopt sophisticated CRM applications and focused instead on minimizing costs.

Another example is offered by businesses that serve mainly transient customers, such as gift shops or food operators adjacent to major tourist attractions (the Colosseum in Rome, for example). A restaurant in such a location has an almost entirely transient customer base. While it will want to provide good service to foster word-of-mouth referrals and, in the age of the Internet, ensure that it gets good reviews, management can extract very limited value from customer data. Such a business should therefore seek to create efficiencies by limiting investments in collecting and using customer data and instead manage each interaction as an individual occurrence.

---

4 Ibid.; Kwortnik, R. J., Piccoli G., and Applegate, L. M. *Carnival Cruise Lines*, Case #806015, Harvard Business School Publishing, 2005; Piccoli, G. "Outrigger Hotels and Resorts: A Case Study," *Communications of the AIS* (15:5), 2005, pp. 102-118; Ives, B., and Piccoli, G. "STA Travel Island: Marketing First Life Travel Services in Second Life," *Communications of the AIS* (20:28), 2007, pp. 429-441; Piccoli, G., Anglada, L., and Watson, R. "Using Information Technology to Improve Customer Service: Evaluating the Impact of Strategic Opportunities," *Journal of Quality Assurance in Hospitality and Tourism* (5:1/2), 2005, pp. 3-26; Piccoli, G., O'Connor, P., Capaccioli, C., and Alvarez, R. "Customer Relationship Management – A Driver for change in the Structure of the US Lodging Industry," *Cornell Hotel and Restaurant Administration Quarterly* (44:4), 2003, pp. 61-73; Piccoli, G., Spalding, B. R., and Ives, B. "The Customer Service Life Cycle: A Framework for Internet Use in Support of Customer Service," *Cornell Hotel and Restaurant Administration Quarterly* (42:3), 2001, pp. 38-45.

---

5 For simplicity, we constrain the focus of this article to direct relationships between a firm and its customers—first-order effects—without addressing the role of second-order effects such as word-of-mouth marketing, the "evangelist" effect, or brand building. Note, however, that being able to measure and harness such second-order effects is predicated on the same attention to data capture and analysis described in this article.

Figure 1: A Four-strategies Framework for Exploiting Customer Data

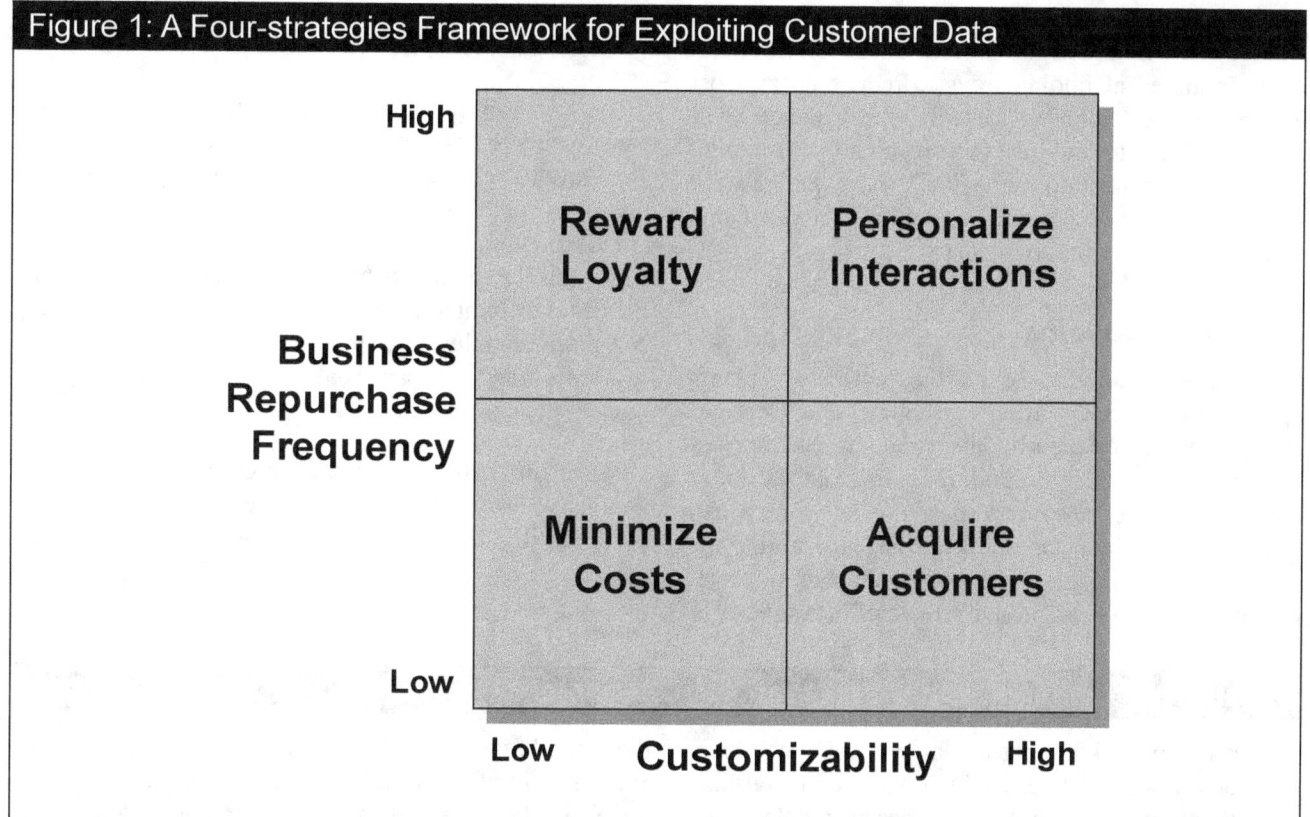

Figure 2: Cost Minimization at OHANA[6]

OHANA Hotels and Resorts, the mid-scale brand of Hawaii-based Outrigger Enterprises Group, is a successful chain of hotels managing seven properties on the Hawaiian island of Oahu and in Guam. Previously branded as Outrigger Properties, these hotels were rebranded in the late 1990s in an effort to bring clear positioning to Outrigger's varied portfolio.

In the words of President and CEO David Carey, "We had huge variation in the portfolio—if you stayed at a budget property vs. a beach-front property, you'd be very confused as to what an Outrigger was." The rebranding makes the value proposition very clear. With average daily rates of just $66, OHANA provides budget-conscious travelers with a clean, consistent, no frills product that allows them to extend their "vacation in paradise" by saving on their accommodation. The firm clearly articulates this notion by positioning itself as "a window on an experience" rather than the experience itself. In other words, OHANA allows people to experience Hawaii by providing the base for their exploration.

While the OHANA product is competitively priced, reaching the Hawaiian islands is still expensive and time-consuming, requiring at least a five-hour flight and an extended stay. As a consequence, OHANA can only expect a relatively low repurchasing frequency and a long repurchase cycle.

By definition, the OHANA product is very simple and minimally customizable. While clean and populated by friendly staff, the typical OHANA hotel is akin to a Holiday Inn, with a limited range of services. As a consequence, when the CRM and personalization craze swept the industry in the early 2000s, OHANA's leadership carefully considered this option but decided against it. The leadership team recognized that the firm would face difficulties in collecting extensive personal and transactional data and that little could be done with the data even if it was collected. Instead of investing in the infrastructure, training, and operation of a customer-data strategy, OHANA opted to maintain its focus on efficient operations and consistent delivery, and to continue using competitive prices as the tool to tip the customer value proposition in its favor.

---

6   For more details, see Piccoli, G., op cit., 2005.

Firms that fall into the Minimize Costs quadrant of the framework will benefit most from data collection and analysis aimed at finding ways to reduce costs. Cost allocation and reporting systems, for instance, will help such firms to fine-tune their revenue-management and price-optimization strategies.[7] The experience of OHANA Hotels and Resorts is representative of this strategy in the hospitality industry.

2. Reward Loyalty

A rewards strategy for exploiting customer data is appropriate when a firm's products and services are purchased frequently, the products or services are fairly standardized, and it is difficult to tailor them to specific customers' needs. Such a firm can use customer data to evaluate the profitability of each customer—actual and potential—and use this information to reward behavior in an effort to increase customer loyalty, create switching costs, or boost its "wallet share."[8]

The firm can also use data collected about individual customers to generate accurate reports and improve its operations (e.g., grocery stores analyzing the contents of a customer's basket). Note that this requires an understanding of individual customers' profitability as well as their propensity to repurchase without incentives, which means the Reward Loyalty strategy is more complex and sophisticated than the "buy one get one free" approach that many firms seem to settle on.

Passenger air transportation is a classic example of an industry in this quadrant. The well-known experience of Harrah's Entertainment is also typical (see Figure 3).

Figure 3: Rewarding Loyalty at Harrah's Entertainment

Harrah's Entertainment is the largest casino company in the United States, in large part thanks to the decision made 10 years ago to proactively focus on rewarding customers.[9] Through its well-known Total Rewards loyalty points program, Harrah's has been able to collect extensive behavioral data about its customers' gambling behavior. Harrah's executives realized very early on that a modern slot machine is a digital computer, capable of recording and transmitting in real time and with complete accuracy all the transactions it performs. Providing customers with a card enables the firm to link a name to these transactions, thereby monitoring behavior over time.

Armed with this infrastructure for collecting customer data, Harrah's is able to extract value from the data. For the average gambler, Harrah's core customer, the gaming product isn't easily customizable. For example, all dedicated slot machine players experience the same feeling of anticipation and excitement when they play, and the firm has little ability to tailor the gaming experience for each of them. On the other hand, avid gamblers play very frequently—often several times a week.

Having deployed its points program, Harrah's is able to reward avid gamblers with dedicated offers as well as allowing them to receive preferential treatment (e.g., skip the buffet line). Privileges such as these create incentives for players to consolidate their gaming at Harrah's, rather than spread their gaming budget over multiple brands. Moreover, Harrah's ability to track players' behavior enables it to proactively measure responses to different types of incentives, thus improving its ability to offer meaningful rewards.

Perhaps the most interesting insight from Harrah's experience is its use of what we call *pre-emptive rewarding*. Because of the inherent characteristics of gambling, particularly slot machine gaming, after accumulating enough transactional data from past guests, Harrah's can quickly and accurately estimate a customer's *future* value within minutes of the player joining the program. The calculation of future value is based on demographic data collected when the customer signs up for the program as well as on transactional data such as rate of play, type of games played, and the like. This enables the casino to start treating the customer according to his or her future value (i.e., to provide pre-emptive rewards) rather than having to wait for observed play before starting to provide rewards.

---

7 For more information, see Cross, R. G. *Revenue Management: Hard-core Tactics for Market Domination* (1st ed.), Broadway Books, 1997.

8 A business's wallet share is a measure of how much customers spend with it rather than with competitors.

9 For more information, see Rajiv, L., and Carrolo, P. M. "Harrah's Entertainment Inc.," Harvard Business School Case (Product Number 9-502-011), 2001.

3. Personalize Interactions

A service-personalization or product-customization strategy is most appropriate for firms where the potential repurchase frequency is high and a high degree of customizability is the norm. Under these conditions, it is possible to collect significant data about individual customers because of the repeated interactions the firm has with returning customers. Moreover, because of the high degree of potential product customization, management has many opportunities to use this data to tailor the product or service to the specific needs—learned or inferred—of returning customers. Thus the firm can use customer data to modify its operations and differentiate its products or services.

The Ritz-Carlton's use of the CLASS system provides a prime example of the Personalize Interactions strategy for exploiting customer data in the hospitality industry[10] (see Figure 4). Another industry that would fit in this quadrant of the framework is event planning, particularly firms that work closely with customers who need many recurring events to be organized (e.g., large investment banks).

4. Acquire Customers

Much conventional thinking about strategies for exploiting customer data seems to imply that such data is not useful for firms with little prospects for repurchase (i.e., repurchase frequency is potentially low). Such myopia misses a key opportunity for firms

---

Figure 4: Personalization at The Ritz-Carlton

The Ritz-Carlton is an international chain of luxury hotels that aspires to provide the "finest personal service and facilities."[11] With its emphasis on tailored service and attention to detail, The Ritz-Carlton has been cited over the years as an exemplar of personalized service.

The Ritz-Carlton, owned by Marriott International, comprises 70 hotels in 23 countries, with 21 more due to open by 2011. Each hotel has access to a centralized database of guest preferences and transactions. The database was originally introduced in 1996, when guest-facing employees carried a pen and notepad on which they would discretely annotate observations about the preferences and interests of guests with whom they came into contact. For example, if a guest called the concierge and mentioned that he wanted only single-malt scotch whiskey available in the bar in his suite, a note would be made. That note would then be input into that hotel's guest database so that single malts would be stocked on the guest's next visit. Other common remarks range from "provide golf balls" to "ensure the room is deep cleaned" to "do not ask guest for her address."

As the chain grew, and it became feasible to create a centralized database of guest preferences, the data from each individual property was made available to all hotels in the chain. Today, employees use a hotline to communicate their observations rather than carry notepads.

To deliver the finest personal service though, it is not enough to anticipate guest needs and provide items guests are likely to request. It is also important that guests' interactions with staff are equally customized and rewarding. For this reason, every day at a Ritz-Carlton hotel begins with a meeting of guest-facing personnel where the "guest recognition daily" is distributed. This document, in the form of a small booklet, lists all the expected guests, their length of stay, the number of previous stays at that hotel, and the number of previous stays with The Ritz-Carlton chain, as well as a list of observed past preferences and comment/action items. The purpose of the guest recognition daily is to enable every employee who is likely to come into contact with guests to be able to appropriately interact with them on a personal basis.

For over a decade, The Ritz-Carlton has consciously positioned its product as a unique, tailored experience. But its ability to do so depends on the fact that its best customers visit its hotels often enough for employees to build deep profiles and preference lists on which they can act. In other words, The Ritz-Carlton's personalization strategy is predicated on a significant repurchase rate and a high degree of customizability that enables its employees to tailor the product to the individual needs and preferences of each guest.

---

10 See Sasser, E. W., Jones T. O., and Klein N. *Ritz-Carlton: Using Information Systems to Better Serve the Customer,* Case #395064, Harvard Business School Publishing, 1994.

11 See "Gold Standards" (http://corporate.ritzcarlton.com/en/About/GoldStandards.htm).

with low repurchase frequency but a high degree of product or service customizability to acquire new customers.

With the Acquire Customer strategy for exploiting customer data, a firm collects exhaustive data about its current customers. This data is analyzed to create customer profiles based on an evaluation of profit margins at the different profit centers where customers transact business, and to develop models to identify and attract new profitable customers while avoiding non-profitable and marginal ones. The availability of such deep business intelligence becomes all the more important during slow periods when marketing budgets get slashed and efficiency in attracting new profitable customers becomes paramount.

The wedding reception business is a good example of a segment that falls in the Acquire Customers quadrant of the framework. Such firms offer highly customizable experiences but typically have low repurchase frequency. Their offerings are also highly seasonal and cannot be scaled (e.g., it is not possible to host simultaneous receptions in one facility). It follows that a successful wedding hall should pay careful attention to the type of customer it attracts. In other words, the optimal strategy for a wedding reception firm is to clearly understand profit margins and customer segments so it can give priority to customers most likely to purchase a bundle of services (e.g., catering, flowers, entertainment) that maximizes its profits. It should use data analytics to build a profile of the most profitable customers and then focus on improving its success rate in recruiting such customers. Carnival Cruise Lines provides an example from the hospitality industry (see Figure 5).

USING THE FRAMEWORK TO DEVISE A COMPREHENSIVE STRATEGY

Most businesses have many types of customers, some recurrent and some infrequent or one-off, and it is therefore hard to unequivocally place a firm into one of the quadrants. But the four-strategy model provides an analytical tool that can help a firm evaluate the advantages and disadvantages of each of the four strategies and, more importantly, the natural fit of each strategy to the firm's characteristics.

Firms should identify which quadrant (or quadrants) of the framework they currently fit into and decide where they want to be. For example, we have used the framework to help a newspaper publisher (low

repurchase frequency, because of its subscription model, and low customizability) rethink its value proposition. The framework helped the company realize that its core asset, a database of new stories, could support on-demand delivery of personalized preselected news categories (e.g., international business news and local sports results) in voice format to a cell phone during the morning or afternoon commute. Thus the firm is morphing its flagship product into an on-demand news delivery service characterized by high repurchase frequency and high customizability. This shift will enable the publisher to exploit a personalization strategy that creates both a superior value proposition and stronger relationships with customers over time.

Note, however, that the positioning of a specific firm within the framework represents a "field of opportunity" rather than the only strategy available to the organization. For example, a firm with significant levels of potential repurchase frequency and highly customizable products can obviously adopt a personalization strategy. Yet, given its ability to collect multiple data points about the same customers, it could also follow a rewards strategy. Conversely, a firm in the Reward Loyalty quadrant cannot adopt a personalization strategy as its products are too constraining to make such a strategy feasible.

Organizations whose field of opportunity spans multiple quadrants of the framework should pay particular attention to prioritizing the four possible strategies for exploiting customer data. For example, a firm in the top right quadrant (high repurchase frequency and high customizability) could conceivably engage in all four strategies. Yet a strong commitment to one of them may be most appropriate.[12]

Firms should pay special attention to potential repurchase frequency when considering their mix of product or service offerings. Resource constraints mean that a business cannot be all things to all people, and designing and launching a new product, service, or initiative usually requires high upfront investments. Firms should therefore look at the potential for payoff over the entire customer base (or a large proportion of it) so that the fixed cost can be spread across a large number of customers. In other words, there is no point in gearing up for personalization (creating the systems and processes, training people, and so on) unless the personalization strategy can yield an adequate return. This will be achieved only if personalization is used by a large number of customers or a limited percentage of

---

12 For more information, see Ghemawat, P. *Commitment: The Dynamic of Strategy*, The Free Press, 1991.

Figure 5: Customer Acquisition at Carnival Cruise Lines[13]

Carnival Cruise Lines is one of the world's dominant cruise operators, carrying over three million passengers in 2007 on its 22 "Fun Ships."[14] With a growing fleet, including many new ships carrying in excess of 3,000 guests and 2,000 crew, Carnival targets its "floating resorts" at almost every demographic. As Carnival's President and CEO Bob Dickinson put it: "We try to position ourselves in the mainstream vacation market. We're the Fun Ships. We're for Everyman with a capital E."

The sheer degree of complexity and multifaceted operations of a modern cruise ship are what enables each guest to have a different experience by defining what they want the "Fun Ship" to be. Guests can spend time—and money—at countless bars, the many on-board shops (including duty-free), the spa, art auctions, wine tastings with the sampled vintages for sale, airbrush art, Internet cafes, video arcades, golf simulators, casinos, shore excursions, ship-to-shore telephones, laundry, and the ship's photo gallery. As well as on-board spending and excursions, guests can chose from various accommodation levels and, before they even sail, can select different routes and ports of call. The huge number of options provides a significant degree of customizability of the cruise product.

However, the potential repurchase frequency of cruise products is low. While there are a few devoted cruisers that produce significant volume, Carnival estimates that only 20% who have cruised with the brand return more than once in the following seven years. The pattern of repeat purchases is shaped like a hockey stick, with low repurchase frequency and long repurchase cycles (e.g., a honeymoon cruise followed by a tenth anniversary voyage). With customers typically very satisfied with the cruise experience, the reasons for low repurchase frequency are structural. Even cheaper cruises are expensive and time-consuming, as they include airfare and transfers, over and above the average $1,651 price tag for a seven-day trip.

Carnival maintains a conservative stance toward new information systems initiatives, which its status as one of the largest and most successful cruise lines enables it to do. However, it has now been collecting and storing voyage detail data for five years. This data is particularly precise as modern cruise ships are cashless; all transactions are completed using magnetic stripe cards. Carnival's "Sail and Sign" card enables precise tracking of all transactions at an individual level. The data collected via the cards therefore enables Carnival to create accurate models of the spending patterns and profitability of past customers. Although the data may be of little value in helping Carnival to increase loyalty and persuade those who have cruised before to return, it should enable the firm to target appropriate prospects. The data can be used to estimate the potential profitability of future customers and therefore enable a conscious customer-acquisition strategy that improves both marketing efficiencies and financial performance.

---

customers who account for the majority of the firm's profits.

LESSONS LEARNED

The cases we have cited, which are all from the hospitality industry, suggest that forward-thinking organizations now pay significant attention to the data that their transaction processing systems generate during the normal course of business. Rather than aggregating or disposing of that data when transactions have been completed, they are increasingly saving it and devising ways to extract value from it. But the cases also illustrate there is a shift away from capturing customer data as a byproduct of transaction processing systems toward a forward-looking data-capture strategy and proactive analysis of the data.

Yet, even these organizations have not fully embraced the born digital approach. Much of the data they use is still produced as a byproduct of existing systems rather than being based on an infrastructure crafted for capturing critical customer data in digital form at its inception.

ADOPTING THE BORN DIGITAL APPROACH

A born digital strategy requires a disciplined approach to digital data capture, predicated on the recognition of the value of customer data and the consequent need to architect systems that originate

---

13 For more details, see Kwortnik, R. J., Piccoli G., and Applegate L. M., op cit., 2005.

14 Source: Cruise Industry News (2007) International Guide to the Cruise Industry, 20th edition.

critical customer data in a digital form. Although we have focused on the hospitality industry, we believe that many other organizations in a wide variety of industries could gain significant benefits from adopting the born digital approach and that the time is now right to do so. Customer service interactions are increasingly computer-mediated both online (e.g., e-commerce transactions) and offline.[15] Every computer-mediated customer service transaction where digital identification of the customer occurs (e.g., a log-in, a swipe card, a customer number) provides an opportunity to capture data. The trend toward computer-mediated customer service is likely to accelerate with the increasing use of mobile technology and the high degree of accessibility and proximity it provides.

The pervasive and affordable network infrastructure ushered in by the Internet provides the second catalyst for the born digital approach. This infrastructure makes it feasible for organizations to develop complete and centralized data stores of their customers' interactions, regardless of the touch point or where in the world they occurred. Organizations are thus creating the prerequisite for profiting from customer information.

Decisions about collecting born digital customer data should be driven by the desire to seamlessly and unobtrusively collect data that answers six basic questions about a transaction:[16]

1. When did the transaction take place (e.g., time and date)?

2. Where did the transaction occur (e.g., store address, touch point)?

3. What was the nature of the transaction (e.g., product purchased)?

4. How was the transaction executed (e.g., company gift card)?

5. Who initiated the transaction (e.g., customer id)?

6. What was the outcome (e.g., value of the sale, satisfaction)?

A good example of the advantages of adopting the born digital approach is provided by the newest multibillion-dollar resort casino in the U.S.—Wynn Las Vegas. Historically, casinos have not placed significant value on customer data. Until the recent publicity given to Harrah's innovative CRM and BI practices, a large casino's standard operating procedure was to value customers on the basis of judgments made by hosts and pit bosses. This unscientific approach tended to give undue weight to the contribution of a few big gamblers—the so-called whales—while under-valuing the multitude of smaller, but often more profitable, players. With the advent of state-of-the-art digital slot machines, it became feasible and cost-effective to build comprehensive profiles of avid slot machine players, without disrupting their play or experience, and craft a targeted rewards strategy.

Casinos are now expanding their use of technology to capture valuable customer data, by embedding radio frequency identification (RFID) transceivers in the chips used at table games. With this approach, a casino can track individual gambling behavior in real time and with the same precision as slot machine usage. Historically, those casinos wishing to place a value on table games players had to use a labor-intensive monitoring system that required pit bosses to note approximate amounts wagered, a process neither simple nor precise. By some estimates, many casinos still provide table games players with incentives that exceed their worth by 20% to 30%.[17]

Embedding RFID transceivers in chips means that table games data is recorded in an easily storable and retrievable format without interfering with the customer's enjoyment. The key insight here is that there are no tangible operational advantages associated with deploying RFID-enabled table game chips. They look and behave like traditional chips, and transactions (i.e., table play) are not affected. The value proposition is rooted purely in the recognition that digital data capture is frequently a prerequisite to the operational and financial feasibility of a comprehensive customer-centric strategy.

Note that in advocating the born digital approach we are not criticizing the historical attention given to the re-use of data created as a byproduct of transaction processing systems. Rather, it is an extension of this tradition. We still urge managers to seek out nuggets of valuable data buried and forgotten in transaction processing systems. But we also advocate that they should take a more proactive approach which ensures valuable data are born digital. The infrastructure and technologies to enable this shift are largely in place. A change in mindset is the biggest obstacle that remains.

---

15 See Froehle, C. M. "Service Personnel, Technology, and Their Interaction in Influencing Customer Satisfaction," *Decision Sciences* (37:1), 2006, pp. 5-38.

16 See Watson, R. T. *Data Management: Databases and Organizations* (5th ed), 2006, Wiley p. 450.

17 Gilbert, A. "Vegas casino bets on RFID," September 6, 2007, http://news.com.com/Vegas+casino+bets+on+RFID/2100-7355_3-5568 288.html.

OBSTACLES AND PITFALLS TO WATCH OUT FOR

A strong word of caution is necessary at this point. While capturing data in digital form does indeed provide a wealth of opportunities to forward-looking firms, several potential obstacles and pitfalls remain. First and foremost, the potential for privacy violations increases tremendously in the born digital world.

In addition to privacy concerns, organizations must manage customer perceptions about their objectives in capturing data in a digital form. The now (in)famous example of RealNetworks' popular RealJukebox software illustrates the dangers.[18] This firm was accused by privacy advocates of surreptitiously collecting information about the CDs customers listened to. Although RealNetworks claimed its intention was to provide better entertainment options for its customers and pointed out that listeners could simply uncheck the tracking option, the perception of misuse and ill-will were all it took for a backlash to occur. Sony BMG made an almost identical mistake in 2005 when it surreptitiously installed a rootkit[19] on the computers of customers purchasing Sony BMG music CDs as part of its approach to Digital Rights Management.

These examples illustrate that, in a world of increasingly available digital customer data, it is imperative that a firm acts transparently in its interactions with customers in order to avoid even the slightest suspicion of misconduct.

Firms also need to consider the impact of any security breaches involving customer data held in a digital form. With the proliferation of digitized customer data, much of it potentially sensitive, a priority for firms that seek to implement data-driven customer strategies must be to develop the procedures, culture, and skills of high-quality customer information stewardship.

ACTION CHECKLIST

We offer the following action checklist for identifying strategic opportunities for exploiting customer data and for adopting the born digital approach:

1. Identify which of the data-driven customer strategies are suitable for your firm by assessing your product or service portfolio in terms of repurchase frequency and customizability.

2. Evaluate the degree of current digital origination of the needed data. You will likely discover that your firm already has the potential to produce significant insights about customer preferences and behaviors.

3. Determine what technologies you need to capture relevant customer data at its inception point and how the investment can be justified. Like many hard-to-justify infrastructure projects, the technology needed to support the born digital approach can often be embedded in specific projects.

4. Continually scan the IT marketplace for new technologies or price declines in existing technology that enable customer-convenient and low-cost digital data capture. Constantly question the state-of-the-art in your industry for capturing data in a digital form in the light of technology improvements and innovations.

5. Pay attention to privacy and security concerns. Ensure that customers know that data is captured in a digital form and understand what it will be used for.

We encourage managers to use the ideas presented in this article to both surface and evaluate potential initiatives. Firms can use the four-strategy framework to bridge the current chasm between the wealth of data generated by the increasingly pervasive computerized infrastructure in many businesses and the lack of guidance that has historically held managers back when it comes to exploiting data.

We all acknowledge, often with a robotic nodding of heads, that data is a valuable resource but do little about it. The born digital era now beginning enables resourceful managers to apply analytic frameworks and software to improve profitability by exploiting customer data.

ABOUT THE AUTHORS

Gabriele Piccoli

Gabriele Piccoli (gpiccoli@uniss.it) is Associate Professor of Information Systems at the University of Sassari (Italy). He is a frequent contributor to *MIS Quarterly Executive* and serves as associate editor at *MIS Quarterly*. His work has appeared in academic and applied journals, and he has recently written the book *Information Systems for Managers: Text and*

---

18 See Macavinta, C. "RealNetworks faced with second privacy suit," http://news.com.com/RealNetworks+faced+with+second+privacy+sui t/2100-1001_3-232766.html.
19 A rootkit is a collection of software that enables administrator-level access to a computer or network.

Cases, published by John Wiley & Sons. His research, teaching, and consulting expertise are on strategic information systems and the use of information systems to enable superior customer service.

Richard T. Watson

Richard Watson (rwatson@terry.uga.edu) is the J. Rex Fuqua Distinguished Chair for Internet Strategy and interim Head of Information Systems at the Terry College of Business, the University of Georgia. He has published in leading journals in several fields as well as authored books on data management and electronic commerce. His current research focuses primarily on Green IS leadership. He has delivered invited seminars in more than 30 countries for companies and universities. He has been president of AIS, co-chair of ICIS, and a senior editor for *MIS Quarterly.* He co-leads the Global Text Project, is a visiting professor at the University of Agder, Norway, and is a consulting editor to John Wiley & Sons.

DEVELOPING AN ENTERPRISE BUSINESS INTELLIGENCE CAPABILITY: THE NORFOLK SOUTHERN JOURNEY[1,2]

Barbara H. Wixom
University of Virginia
(U.S.)

Hugh J. Watson
University of Georgia
(U.S.)

Tom Werner
Norfolk Southern
Corporation (U.S.)

Executive Summary

Although many IT and business managers today may be lured into business intelligence (BI) investments by the promise of predictive analytics and emerging BI trends, creating an enterprise-wide BI capability is a journey that takes time. This article describes Norfolk Southern Railway's BI journey, which began in the early 1990s with departmental reporting, evolved into data warehousing and analytic applications, and has resulted in a company that today uses BI to support corporate strategy. We describe how BI at Norfolk Southern evolved over several decades, with the company developing strong BI foundations and an effective enterprise-wide BI capability. We also identify the practices that kept the BI journey "on track." These practices can be used by other IT and business leaders as they plan and develop BI capabilities in their own organizations.

THE IMPORTANCE OF BUSINESS INTELLIGENCE

There is considerable evidence of the importance of business intelligence (BI) for organizations. The origins of BI stem from decision support systems, which first emerged in the early 1970s when managers used computer applications to model business decisions. Over the years, other applications, such as executive information systems (EIS), online analytical processing (OLAP) and data mining/ predictive analytics became important. Today, BI is a broad category of technologies, applications, and processes for gathering, storing, accessing, and analyzing data to help its users make better decisions.[3]

BI has been at the top of CIOs' strategic agendas for the past several years, according to Gartner research,[4] and Forrester expects the BI market to grow from $8.5 billion in 2008 to $12 billion in 2014. Consulting firms are launching new practices and analytics centers to address the increased demand for BI services, and books that illustrate the value of BI and analytics to organizations are best sellers.[5] Fueling the growth in BI are exciting technology advances in the areas of social computing, unstructured data analytics, mobile delivery, and "big data"—datasets that grow so large that they become awkward to work with using conventional database management tools. However, building an enterprise BI capability does not occur overnight. Instead, it is a journey during which foundational competencies are developed over long periods of time.

MISQE is
Sponsored by

KELLEY SCHOOL
OF BUSINESS
INDIANA UNIVERSITY

Delivering Business Value
Through IT Leadership

---

1 Jack Rockart is the accepting senior editor for this article.

2 The authors gratefully acknowledge important contributions from Professor Jeff Hoffer of the University of Dayton; Linda Richardson, Senior Designer Information Systems, Norfolk Southern Corporation; and the many study participants who work at Norfolk Southern. The authors would like to thank Teradata for financially supporting this case study. We also are grateful for partial funding from the Von Thelan Fund at the University of Virginia's McIntire School of Commerce.

3 An excellent overview by Dan Power of the history and evolution of decision support systems may be found at: http://dssresources.com/history/dsshistory.html. See also Turban, E., Sharda, R. and Delen, D. *Decision Support and Business Intelligence Systems,* 9th Edition, Prentice Hall, 2011.

4 See Gartner's 2011 CIO survey results: http://www.gartner.com/it/page.jsp?id=1526414.

5 For example, in 2010 Deloitte launched Deloitte Analytics; SAS, Teradata and Elder Research launched the Business Analytics Innovation Center; and Accenture opened an Analytics Innovation Center. Recent bestsellers include Davenport, T., Harris, J., and Morison, R. *Analytics at Work: Smarter Decisions, Better Results,* Harvard Business Press, 2010; and May, T. *The New Know: Innovation Powered by Analytics,* John Wiley and Sons, 2009.

In this article, we describe Norfolk Southern Railway's[6] BI journey. We interviewed 30 business and IT leaders from across Norfolk Southern and reviewed archival documents produced by the data warehouse team to understand how BI has evolved over the last two decades.

For most of its history, Norfolk Southern operated in a predictable, regulated environment, and the company's business strategy focused on efficiently transporting freight from one point to another. With deregulation in the 1980s, railroads were free to compete in new ways. Deregulation also opened up the possibility of mergers and acquisitions. At the end of the twentieth century, Norfolk Southern acquired 58% of Conrail and developed into a service-oriented, scheduled railroad. The success of this strategy required significant investment in BI, including increased use of optimization algorithms, improved performance monitoring systems, and providing customers with access to up-to-date shipment data. Today, the railroad focuses on customer service, fuel efficiency, asset usage, and workforce productivity, and it leverages BI across the enterprise to help accomplish this.

The Norfolk Southern case provides an excellent example of how the BI journey occurred in one organization. From our analysis of Norfolk Southern's experience, we provide an organizing framework for enterprise BI and the key practices that other organizations can use to help them "stay on track" during their own BI journeys.

NORFOLK SOUTHERN'S BI JOURNEY

Norfolk Southern Railway Company is one of the four largest Class I railroads in the U.S. The railway includes approximately 20,000 route miles in 22 eastern states and the District of Columbia, serves all major eastern ports, and connects with rail partners in the West and Canada, linking customers to markets around the world. Norfolk Southern provides comprehensive logistics services and offers the most extensive intermodal[7] network in the East.

---

6 Norfolk Southern, based in Norfolk, VA, is a Class I major freight railroad. There are over 1,000 railroads in the United States, of which, just seven are classified as Class I.

7 Intermodal refers to a transportation route that involves more than one mode of transportation. Norfolk Southern's Intermodal Division handles routes that combine shipping and trucking services along with rail. Thus the company can, for example, offer customers a service that originates at a manufacturing plant and ends at a retail location, using various modes of transportation along the way.

Competing in a Regulated Industry

Just a few decades ago, the U.S. railroad industry was highly regulated. Like most railroads, Norfolk Southern focused on moving shipments from point A to point B, and in doing so, the company provided a good cost advantage over other types of transportation, such as trucking. Consistent with this business approach, Norfolk Southern had implemented complex transaction-based information systems to support the movement of rail cars from point A to point B safely and efficiently. These systems were central to the operations of the railroad.

Unfortunately, these systems could not be used for any significant reporting purposes because processing queries might well degrade system performance. When employees needed a report, they had to submit a request to the IT organization; the resulting report would take time to produce and require considerable IT resources. Therefore, access to reporting facilities, particularly for specialized reports, was limited.

Post-deregulation, however, business functions at Norfolk Southern began thinking in new ways and had to make new kinds of decisions concerned with pricing and cost management. By 1995, employees within the Marketing and Cost Departments had developed enough reporting needs to justify investment in a dedicated reporting system. Together, these departments funded a one-terabyte data mart, which was updated each night with records of all the rail car movements from that day (e.g., arrivals, departures). This initial BI effort supported basic reporting about customer service and cost data to help the company understand "How was Norfolk Southern serving its customers?" and "What should Norfolk Southern charge its customers based on the cost of moving goods from one point to another?"

At first, Marketing Department employees with IT skills implemented and maintained the data mart, and a contractor developed the data models for the system. However, significant investment was required for the technology platform to support the data mart, and the skills of IT professionals were required to maintain it. Thus, when a member of the Marketing Department involved with the data mart transferred to the IT Department, the company moved the responsibility for the data mart to the IT Department.

At this point, the system still remained oriented toward the Marketing and Cost Departments, but the platform was housed and managed by corporate IT, and the data mart began to resemble an enterprise data warehouse. This was a critical step in the evolution of BI at Norfolk Southern. This move not only positioned BI to become available across the enterprise but also allowed the IT Department to create proper controls and apply standard IT practices (e.g., data standards) to the warehouse so that it could become a sustainable technology platform. The original data modeling contractor continued working with the data warehouse after the move and helped the IT Department to effectively build on what had been achieved by the Marketing and Cost Departments.

Becoming a Scheduled Railroad

In 1999, senior management at Norfolk Southern invested in a growth strategy by acquiring 58% of Conrail. This move, known as the "Conrail Split," increased Norfolk Southern's size by 50% while providing direct track lines to the New York and Philadelphia markets and ownership of expanded intermodal capabilities. As CEO Wick Moorman explained, this acquisition prompted a two-fold effort:

> *"First, we needed to come up with a new operating plan. Second, we needed to put in place underlying systems and information tools to support the maintenance and the management of the plan."*

The result was an initiative in 2002 called the Thoroughbred Operating Plan, or TOP, to redesign Norfolk Southern's operations. This was a fundamental change for the company. Prior to TOP, the primary operational objective was to maximize train size to minimize the number of train crews needed. Crews were the major variable, and the most visible cost to the railroad. As a consequence a car could stand idle for a day or so and in turn impact other trains to which it would be connected. Scheduled delivery dates sometimes varied within a window of up to three to five days. At the time, this is the way most railroads operated.

With TOP, management invested in new transaction-based information systems and processes that used operations research techniques to determine when and how rail cars should move throughout the Norfolk Southern transportation network. The new systems optimized inventory management and trip planning.

Once Norfolk Southern had implemented the systems to optimize the operating plan for its rail cars, employees throughout the organization needed measures, reports, and tools that would help them manage to the plan. Field managers needed to monitor performance and identify the root causes of going off-plan so that they could make adjustments.

The more field managers conformed to plan, the faster equipment would move through the system, arriving on time more often and spending less time in terminals. As a consequence, a TOP steering committee of senior-level managers funded a new BI data warehouse application that would provide managers with the information needed to manage to the TOP plan.

The TOP BI application analyzed trip plans (i.e., itineraries) for every shipment and determined which trains would handle the shipment and how, when, and where connections between the trains would be made. This application then provided field managers who were accountable for sticking to the TOP plan with graphical depictions of actual performance against the trip plan for both train performance and connection performance (see Figure 1 for a screenshot of the TOP BI application). This operational BI application is a performance dashboard, and its use is embedded within important operational processes at Norfolk Southern.

Over time, Norfolk Southern has reinforced TOP through incentives; for example, a portion of corporate bonuses is tied to how well the railroad runs to plan. In the eight years since TOP was implemented in 2003, Norfolk Southern has reduced rail car cycle time by one day, which translates into millions of dollars of annual savings.

Using accessNS to Compete on Service

With TOP, service became predictable and Norfolk Southern strengthened its ability to compete on service. Becoming a service-oriented, scheduled railroad created huge opportunities. The Marketing Department initially provided visibility into the company's extensive transportation network through its investments in BI reporting. Then, as Norfolk Southern became more "scheduled," it expanded services to customers using a BI application called accessNS.

Customers want to know where their shipments are "right now"—and at times, they also want historical information: Where did my shipment come from? How long did it take to arrive? What were the problems along the route? Prior to the early 2000s, customers would call a Norfolk Southern customer service agent with questions and then wait for minutes, hours, or days for an answer. Behind the scenes, agents had to either ask the IT Department to provide the requested information or navigate legacy systems that were hard to use.

Figure 1: TOP Business Intelligence Application Screenshot

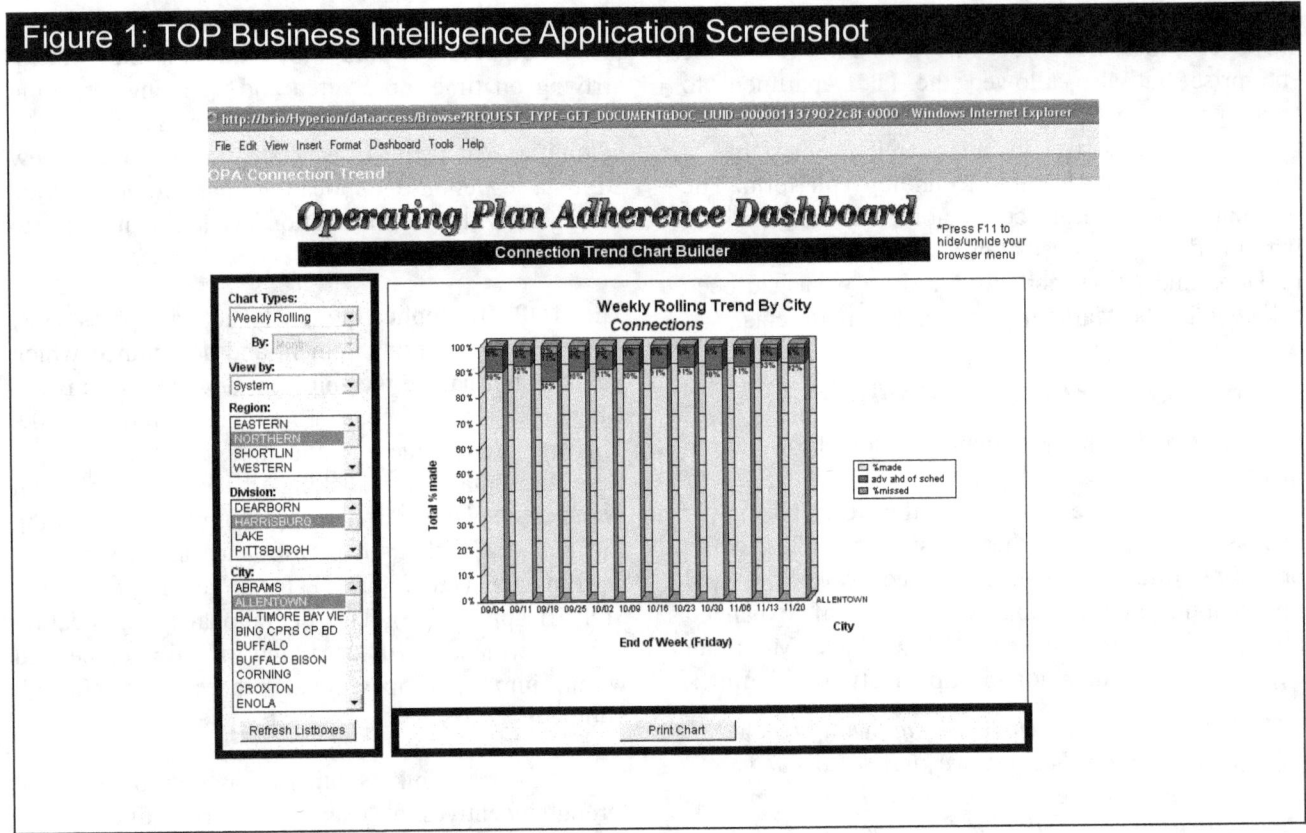

The accessNS application allows customers to access BI reports from the Internet and find answers to questions about service status and performance. This customer-facing BI application was the first of its kind in the industry and was immediately popular with customers. Over time, the user base grew to more than 11,000 in 8,000 customer organizations. Users log on to accessNS to access any of the 20 standard reports and retrieve information at any time. They can access current data, which is updated hourly, or they can look at three years of history. The application uses push technologies to provide alerts, text messages, and RSS feeds; 4,500 reports are now pushed to users daily.

The standard reports can meet a wide variety of user needs. Sometimes, though, customers have a one-off request, such as quickly needing a piece of information for a meeting, or they want information in a different format. For these cases, Norfolk Southern offers a Report Wizard that allows customers to manipulate over 125 data fields to modify existing reports or build new ones (see Figure 2 for Report Wizard screen shots). This capability makes it possible to access information without writing complex SQL queries. Users employ a drag and drop interface (e.g., for sorting fields, ordering columns, specifying limits) to build queries, and the Report Wizard automatically translates the queries into program code. In the words of Blair Hanna, Manager eCommerce:

"The users love it. It takes most business users only 5 to 10 minutes to customize their reports, and sometimes new users never need to contact us at all while creating their own reports."

The success of accessNS had a positive effect on Norfolk Southern's organizational capabilities. Deborah Butler, the company's CIO, explained:

"The customer service center became a real customer center instead of a traditional call center. It became a lot smaller, with much more focus on finding areas where service was not meeting expectations—and finding them before our customers brought them to our attention."

The BI team and business sponsors proactively worked to make internal and external users of accessNS and other BI applications self-sufficient and, in doing so, invested significant time and effort in interface design. Norfolk Southern developers built BI applications with point-and-click interfaces for querying and with dashboards for monitoring. They built custom wizards and web-based portals. They made screens graphically and visually straightforward. The intent was to create intuitive interfaces that users can understand immediately. One application developer explained:

Figure 2: accessNS Report Wizard Screenshots

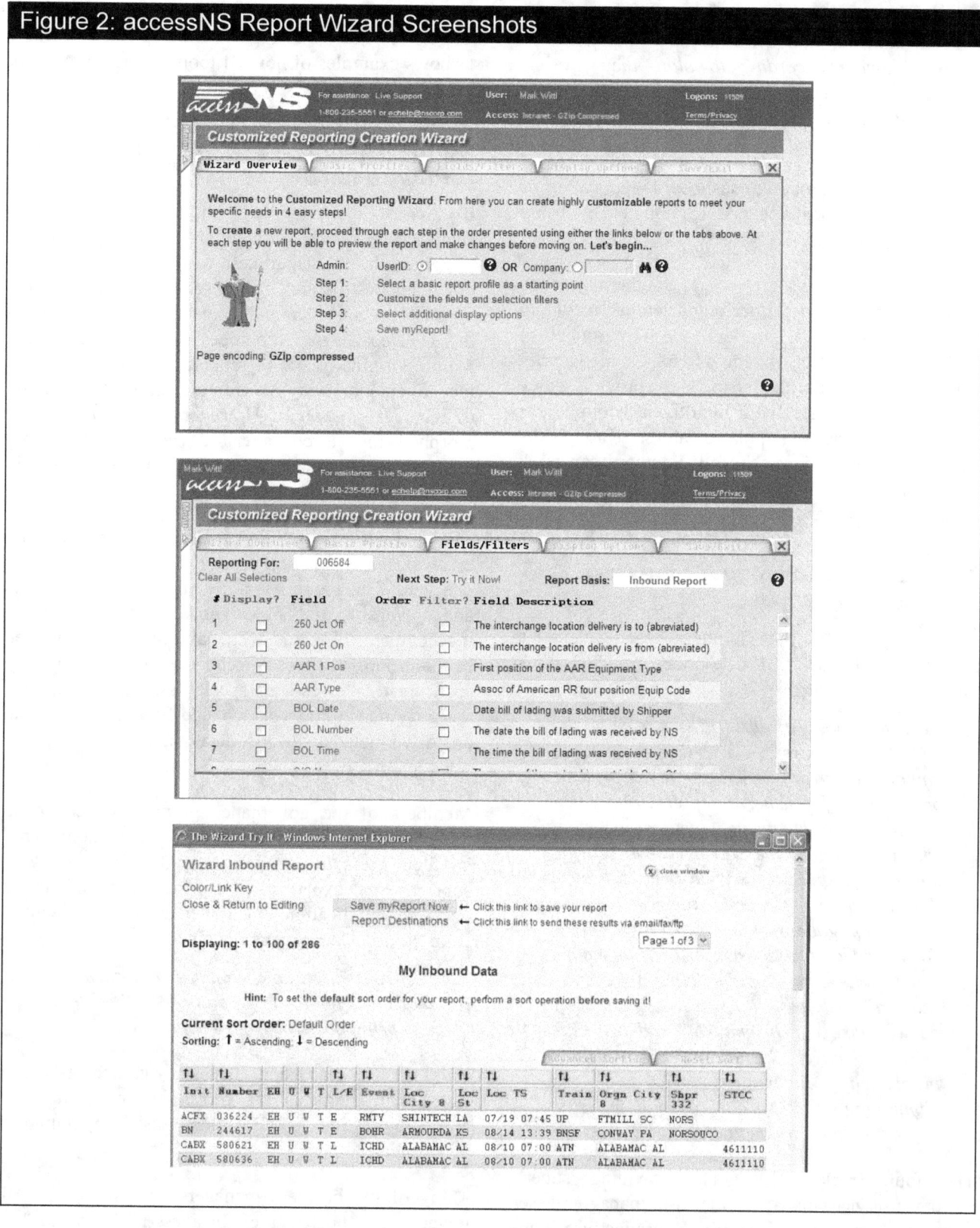

"If I have to put together an expensive training program, then my tool isn't simple enough to use. Our goal is for end users to click on a link or push a button."

Additionally, Norfolk Southern has found that the needs of BI application users change over time. Initially, business users wanted applications that provided visibility into operations, and applications were created to do that. But, according to an application developer, that need has changed:

"Two to three years later, the business users said that they didn't need to know if freight was moving according to plan—they just wanted to see exceptions. So we moved toward exception-based reporting where we clearly highlight problems."

One new exception-based dashboard application is refreshed automatically every 15 minutes by using AJAX technologies.[8]

The external BI users at Norfolk Southern grew to include suppliers and partner organizations, including the U.S. Government. In fact, the Department of Defense and the Department of Homeland Security require transportation companies to provide quick responses to inquiries about high-threat commodities traveling in heavily populated areas. Through its existing BI applications, Norfolk Southern can track goods easily and quickly, enhancing the ability for emergency response.

Track 2012: A Strategy for the New Decade

In recent years, Norfolk Southern has developed a strategy called Track 2012, described by CEO Wick Moorman as follows:

"The first goal of Track 2012 is service. At the end of the day, we are a service business ... and our company will rise and fall depending on the level of transportation service we offer. Second, we need to manage our cost structure, which determines the value we can provide and the margins we can earn. We focus on three big drivers. The first is fuel use. We burn 500 million gallons of diesel fuel a year; how do we reduce fuel consumption? The second is asset turns, or productivity. We are an extraordinarily asset-intensive business. How do we make our locomotives and rail cars more productive? The last driver is workforce productivity. Running to schedule and eliminating variations has a helpful impact in improving labor productivity."

The four Track 2012 goals—providing superb customer service, managing fuel use, managing asset turns, and improving workforce productivity—are enabled by the variety of BI tools used by departments across Norfolk Southern, ranging from Accounting and Human Resources to Operations and Fleet

Management to Strategic Planning. The BI tools range from queries to reports and applications. Table 1 shows examples of how BI tools are tied into Track 2012 goals.

Norfolk Southern's Governance Structure for BI

By August 2007, the large number of BI business sponsors located throughout Norfolk Southern prompted the IT Department to institute a formal BI governance structure. This structure brings together 26 business managers and three representatives from the IT Department who meet for two hours every month. Originally, the IT representatives spearheaded the governance body to promote more effective use of technical resources. However, the group quickly morphed into a customer-focused body motivated to promote BI as an enterprise resource at Norfolk Southern.

The group is chaired by a business manager, and every business function represented in the group has one vote. The group has a set of founding principles and a website, and annual updates are provided to the company directors who have members involved in the group. At times, special interest groups (SIGs) are formed to investigate special issues. For example, the governance body has created SIGs to investigate and drive metadata efforts, data quality issues, and the selection of a next-generation BI tool for the company.

Members of the governance group have expanded their understanding of BI through information shared by their peers in the business. At each meeting, a subject matter expert makes a presentation on existing data or an application. One governance group member from the Tax Department explained:

"This has been an eye opener. I've seen data that I was not aware of and capabilities that I can bring back to my own department."

This Tax Department manager learned how to use e-mail alerts in tax processing when he saw a demonstration of BI alerts used by another department.

Many of the business managers who sit on the BI governance board are business-IT "hybrids"—employees with an impressive mix of business and IT knowledge. In general, employees at Norfolk Southern have long tenures and often move between departments. Over time, the company has developed hybrid employees as people moved from IT into the business, and *vice versa*, as people were specifically

---

8 AJAX (Asynchronous Javascript and XML) is a group of interrelated web development methods used on the client side to create interactive web applications.

Table 1: Track 2012 Goals and BI Applications		
Track 2012 Goal	**Department**	**BI Tools**
Providing Superb Customer Service	eCommerce	accessNS
	Intermodal	Intermodal Operational Dashboard: A real-time (updated every 15 minutes) exception-based dashboard that communicates operational information about intermodal services.
Managing Fuel Use	Industrial Engineering	An ad hoc application was developed after Hurricane Katrina to ensure that fuel was delivered to the right places at the right times in an emergency situation.
Managing Asset Turns	Modalgistics	The Multi-level application optimizes the movement of special multi-level cars.
	Rail Car Distribution	The Empty application reduces the number of empty rail cars that travel on the track network.
Improving Workforce Productivity	Crew Call	The Crew Call application optimizes crew scheduling to ensure crews are at the right train at the right time. It has resulted in $2.8 million annual savings by reducing the cost of trains held for crews.
	Human Resources	The Workforce Planning application predicts future staffing needs. HR planning personnel proactively examine departmental staffing and explore historical trends for retirement ages and years of service. Armed with this data, they forecast retirement attrition for the next 5 to 10 years and help put strategies in place to meet hiring needs.

hired because they possessed a unique combination of both technical and business skills and as people worked in specialty BI reporting groups within business units. As a result, Norfolk Southern currently has a large base of workers who are well equipped to translate business requirements into actions enabled by BI. A text box below provides examples of these four different ways in which Norfolk Southern has built business-IT hybrids.

AN ORGANIZING FRAMEWORK FOR BI AND ENABLING PRACTICES

At Norfolk Southern, BI began with a data mart reporting solution for the Marketing and Cost Departments. Today, it directly supports corporate strategy and is used pervasively across the enterprise, as evidenced by the following three quotes:

> *"Everybody across Norfolk Southern is viewing the same set of facts. It is very hard to put a numerical value on that. Having an enterprise view of data is hugely important and probably one of the most important benefits that has come from our BI program."* (Deborah Butler, CIO)

> *"When someone walks down the hall or our executive department at Norfolk [has] a question about something, within minutes, we can be looking at data and analyzing whatever the situation is. We have all kinds of data now that is out there—projections, runs, timed queries that tell us the status of the railroad at the push of a button—all because of this data."* (Andy Fitzgerald, Manager Car Service)

> *"There will always be questions that we weren't expecting at the time we were putting together the data warehouse. With an enterprise view, you certainly have all the information that you need. You don't want to put stuff in the data warehouse just for the sake of it but it certainly makes doing analysis a whole lot easier. One of the things that I was asked by our Vice Chairman this morning was: 'What's been the impact of the floods in the Midwest on our operation?' Having an enterprise view with that breadth of information makes it possible for me to say, 'Yes, we can get that information without any problem.'"* (Fred Ehlers, VP Customer Service)

It took years for Norfolk Southern to build its BI capability, and Table 2 describes the highlights of the

---

How Norfolk Southern Builds Business-IT Hybrids

Moving from the IT Department to a Business Unit

When one employee started with Norfolk Southern's IT Department, he was assigned to work with financial systems. Over the years, the financial systems evolved from being very clerical to being highly automated. To manage and operate these systems, it is necessary to understand both accounting and technology. This employee is now in the Accounting Department, where his strong technology background is very useful.

Moving from a Business Unit to the IT Department

Another employee took a more circuitous route from the business to IT. He began his Norfolk Southern career in the Engineering Department. He then moved to Accounting and later spent 12 years in Transportation in the customer service area, where he worked extensively with BI as a user. In 2005, he was asked to join IT and lead the BI tool group. This employee explains, *"The idea is to take someone from the business side with years of business experience and with experience as a BI user and bring them into IT to work in the business intelligence area. I had been using the BI tool for many years, and I also understood the business and knew a lot of people."* His vast knowledge of the business helps the IT Department to better understand the needs of the business, and he enjoys tremendous credibility among business users.

Hiring People with a Combination of Technical and Business Skills

The Equipment Planning Department has five equipment managers who are responsible for different types of rail cars—gondolas and flat cars, boxcars, covered hoppers, automotive multi-levels, and auto parts. All of these managers came into the department with strong analytical backgrounds, and all have college degrees: three are mechanical engineers, and one has a master's degree in finance and business. They all know how to access data, analyze it, and get the most out of the information. They are self-sufficient in terms of using the data warehouse.

People Working in Specialty BI Reporting Groups within Business Units

Norfolk Southern has several functional departments that have created BI reporting groups. These groups are not considered as shadow IT groups. Instead, they liaison with the IT Department, working on requirements and report development. A member of one of these groups explains, *"My skill set is pretty technical, but I am exposed to the business side of things because I'm working with the business folks on the same floor. I'm also familiar with IT and know a lot of folks in IT because we stay close and work together. My group is the innermost link between the business and IT. We make sure requirements are 100% correct. And we make sure IT technically delivers in the right way."* The BI reporting groups use iterative development and prototyping to create BI reports and applications that are ultimately passed to the corporate IT function to implement.

---

company's journey from the perspectives of business strategy, data, and BI tools.

Norfolk Southern's journey as described in Table 2 has resulted in a BI capability that can be represented by the organizing framework depicted in Figure 3. First, the company created the critical data foundations required for BI—high-quality, integrated enterprise data that is usable by business users. Second, Norfolk Southern's business functions identified and communicated important strategic business requirements to the IT Department, enabled primarily by the company's business-IT hybrid employees and a strong governance structure. Third, Norfolk Southern has a variety of BI tools, ranging from queries to reports to applications that exploit the data foundations to enable business strategies from across the organization. Finally, the BI capability produces business value that varies from bottom-

line cost savings and revenue generation to improved business processes and better decision making.

Keeping the components of this framework aligned over time is easier said than done. We have identified six practices applied by Norfolk Southern that have helped the company keep its BI initiatives "on track."

Practice 1: Create a Business-run BI Governance Structure that is Meaningful to the Business

BI governance should include a group of engaged and committed business sponsors from across the company. The governance structure at Norfolk Southern enables business functions to share information and spread BI practices. It also allows the business groups to voice opinions about BI, alter BI priorities, and provide input into BI decisions. The challenge, however, is to keep the attention of a large

Table 2: Norfolk Southern's Journey in Building its BI Capability

Strategic Journey	Data Journey	BI Tools
New kinds of decisions on pricing and cost had to be made after the industry was deregulated.	Marketing and Cost Departments funded a one-terabyte data mart to house rail car data, which was loaded nightly.	Marketing and Cost employees used basic reporting and ad hoc query tools.
Acquisition of a majority share in Conrail prompted investments in new operating systems that optimized rail car movement.	Responsibility for the data mart was moved to the IT Department and became a data warehouse that supported functions across the company; enterprise data models, data standards, and controls were developed.	Performance management dashboard applications were built to support decision making in the field; intuitive user interfaces were created.
Norfolk Southern strategically differentiated itself from competitors through its service.	More subject areas of data were included in the data warehouse and update frequency was increased from nightly to hourly to better support customer queries.	A customer-facing web portal BI application was built for customers to provide pre-formatted reports and ad hoc query capabilities; reporting became more exception-based, using push technologies, such as alerts and RSS feeds; focus was on creating decision support applications with short learning curves.
Top management created a new corporate strategy that focused the organization on service, fuel use, asset turns, and workforce productivity.	More subject areas of data were added along with new data types, such as geospatial data; an enterprise governance board initiated efforts to ramp up the areas of data quality and metadata.	A diverse portfolio of BI tools and applications now exists across the enterprise, leveraged by both internal and external users; the user base has grown to include a large number of business-IT hybrid employees.

Figure 3: BI Capability Framework

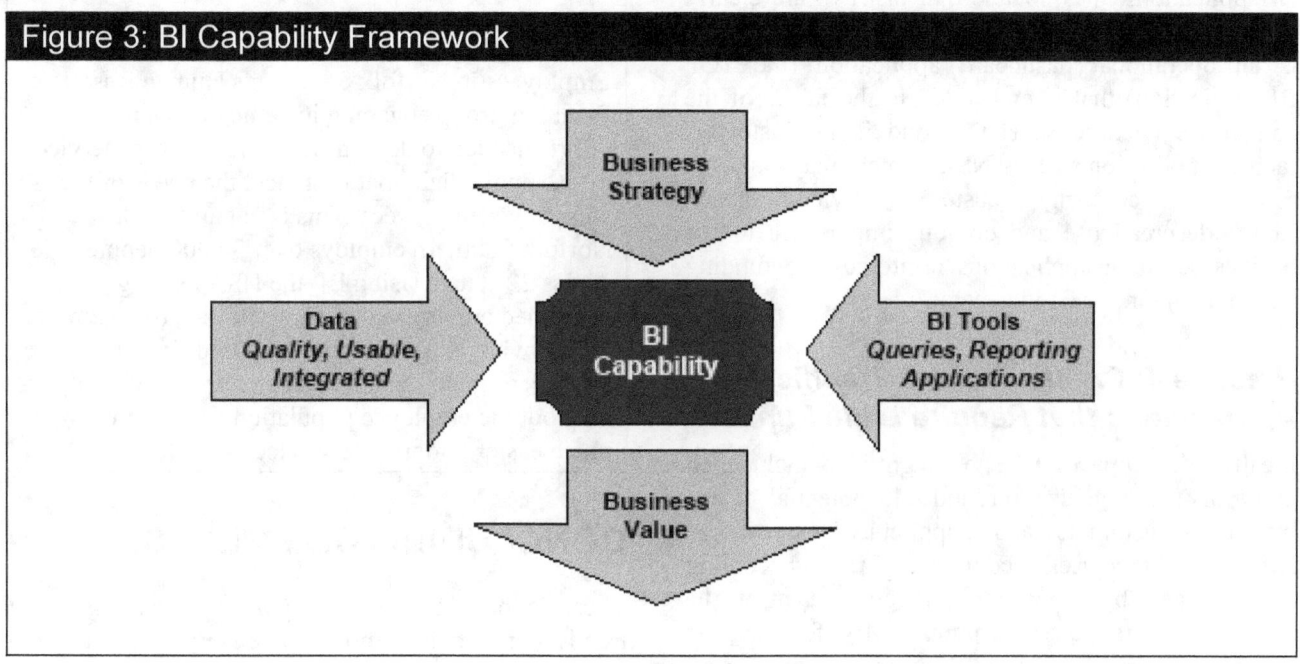

number of sponsors and to leverage their support in productive ways. Attendance at governance meetings is treated very seriously at Norfolk Southern. In fact, members of the group are not allowed to delegate their seats to anyone else. All of this has helped deepen BI sponsorship, creating a strong foundation for enterprise commitment.

Practice 2: Develop Business-IT Hybrids Who Understand Real Business Needs

BI delivers value when reporting and analysis capabilities are highly aligned with business requirements. In many organizations, however, there is a chasm between IT people who build applications and the people who use them. Differences in the domain knowledge and application development skills of these two groups can result in lengthy application development cycles and applications that don't meet user needs. Norfolk Southern bridges this chasm through hybrid employees who have both domain knowledge and the technical skills to develop applications. Organizations should encourage rotational practices, hiring practices, and BI competency groups that align BI reporting and analysis with user requirements. Ultimately, as BI reporting and analysis grows in relevance and importance, so does enterprise usage of BI.

Practice 3: Choose Your BI Applications Wisely

Although Norfolk Southern has a large portfolio of BI applications, it is notable that just two are clearly at the top of the company's priority list. The first is an operational dashboard application (the TOP BI application) that is embedded in the fabric of the company's operations. The second is its customer-facing application (accessNS), which plays a core role in the company's customer service strategy. The widespread use and obvious importance to the business of both applications reinforces the funding and momentum for BI in general.

Practice 4: Create Usable, Graphical Applications that Require Little Support

Ideally, the scope of BI should grow to include all parts of the organization and all potential users, including customers and suppliers. However, as BI scope increases, companies risk becoming overwhelmed by having to support and meet the needs of a large user community. By focusing on creating self-sufficient users through providing easy-to-learn graphical interfaces (dashboards) and through

push technologies (e.g., alerts, RSS feeds), Norfolk Southern has reduced support and training costs, while meeting user needs more effectively. This encourages more and more users to embrace BI because the learning curve is not intimidating.

Practice 5: Don't Ignore the Basics of Data Management

Norfolk Southern has learned that you cannot ignore the basics of the data that support BI—data quality, standards, and metadata. In some respects, the company was fortunate to have a single contractor who has modeled the BI data since the inception of the warehouse and has created standards and maintained consistency as new data types were added.

Practice 6: Treat Data as an Enterprise Asset

Even with quality data, standards, and metadata, companies need to have a culture that supports an enterprise view of data. The governance group at Norfolk Southern reinforces data as an enterprise asset and spearheads special data initiatives, such as metadata and data quality projects, when gaps are identified. Additionally, the company has an open data philosophy whereby user groups are encouraged to share data. This leads to new and interesting uses of data, an example of which is described in the text box below.

Example of the Benefits of Sharing Data Across Departments

Employees in Norfolk Southern's Human Resource Planning group engaging in strategic planning efforts needed to determine where to locate service offices in the field that best meet the needs of the company's employees. This is not an easy task: Norfolk Southern employs over 30,000 people across 22 states. Using BI, the HR Planning group combined employee demographic data (e.g., zip codes) with geospatial data used traditionally by the Engineering Department and was able to visually map out the employee population density, making it much easier to optimize service office locations.

CONCLUDING COMMENTS

There is now a bewildering array of emerging BI trends—big data, mobile delivery, unstructured data mining, social analytics, collaborative BI, and many others. The danger is that companies might be tempted to jump on board the latest bandwagon in the belief that it will provide instant solutions to all their BI requirements. However, a BI capability—on an enterprise scale—cannot be created overnight: most companies with a true enterprise BI capability have taken years to acquire their BI staff, build a comprehensive data infrastructure, select and implement BI software, develop applications, train users, and more. Norfolk Southern is an example of a company whose BI journey has remained "on track." We believe the practices that helped Norfolk Southern accomplish this provide guidelines for other organizations as they progress in their BI journeys.

ABOUT THE AUTHORS

Barbara H. Wixom

Barbara Wixom (bwixom@virginia.edu) is an associate professor and director of the M.S. in the Management of IT program at the University of Virginia's McIntire School of Commerce. She is an associate editor of *The Business Intelligence Journal*, a fellow of The Data Warehousing Institute, and an instructor in data warehousing, database, and strategy at undergraduate and graduate levels. She has published in journals that include *Information Systems Research, MIS Quarterly,* and the *Journal of MIS,* and has two textbooks published by John Wiley & Sons.

Hugh J. Watson

Hugh Watson (hwatson@terry.uga.edu) is a professor of MIS and a holder of a C. Herman and Mary Virginia Terry Chair of Business Administration in the Terry College of Business at the University of Georgia. He is a leading scholar and authority on business intelligence and analytics, having authored 22 books and over 150 scholarly journal articles. Watson is a fellow of the Association for Information Systems and The Data Warehousing Institute, and is the senior editor of the *Business Intelligence Journal.* For the past 20 years, he has been the consulting editor for John Wiley & Sons' MIS series.

Tom Werner

Tom Werner (Tom.Werner@nscorp.com) is Vice President of Information Technology at Norfolk Southern Corporation. He holds a B.S.E. in Electrical Engineering and Computer Science from Princeton University and an M.B.A. in Finance from the Wharton School of the University of Pennsylvania. Prior to joining Norfolk Southern in 1999, he held consultant and management positions in Arthur Andersen/Andersen Consulting, Bankers Trust, and KMPG.

Harvesting External Data: The Potential of Digital Data Streams

Events, individuals' experiences and actions are increasingly "born digital," captured in real time by ubiquitous sensor networks. These events, experiences and actions are ready-to-process data that is available in continuous streams that are constantly evolving. We explain how to acquire, understand and use real-time digitally generated data for new value-creation opportunities that require new capabilities to be implemented.[1]

Gabriele Piccoli
Università di Pavia
(Italy)

Federico Pigni
Grenoble Ecole de Management
(France)

The Growing Opportunities Provided by Digital Data Streams

We live in a world where human activity is monitored and recorded, both in terms of physical activities and—even more so—digital activities. Streams of digital data are being created in massive quantities by numerous sensors—data that can be used both for real-time tactics and long-term strategy. In this article, we explain how organizations can leverage these *digital data streams* (DDSs) to increase consumer value and to improve operational efficiency. Much of these data streams flow from consumers using the Internet, but DDSs are generated also from, for example, cars with a speed-pass device driving through a toll booth, personal activity monitored by digital cameras or even a sensor for soil moisture. The challenge is first to understand the potential of DDSs and then take advantage of the opportunities they provide.

Before discussing the elements of DDSs, it's important to distinguish them from "big data," which concerns the vast amount of data available today that stretches the limits of traditional database architectures.[2] Though DDSs are related to big data, they have a different nature.

---

1   This article is based on research sponsored by the Advanced Practices Council of SIM.
2   See: Beyer, M. A., Lapkin, A., Gall, N., Feinberg, D. and Sribar, V. T. *'Big Data' Is Only the Beginning of Extreme Information Management,* Gartner, April 2011; Manyika, J., Chui, M., Brown, B., Bughin, J., Dobbs, R., Roxburgh, C. and Hung Byers, A. *Big data: The next frontier for innovation, competition, and productivity,* McKinsey Global Institute, May 2011.

Figure 1: Three Stages of a Digital Data Stream

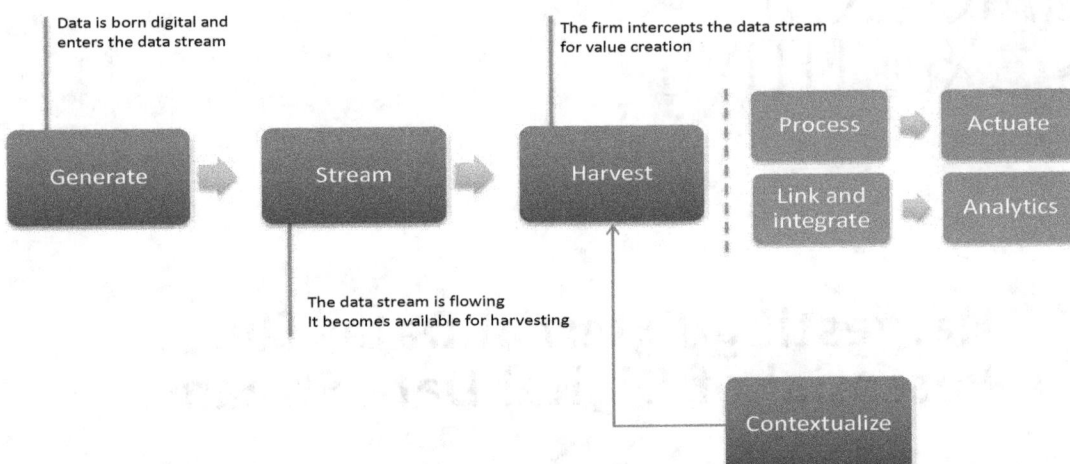

Big data involves mountains of generally static data that can be mined for insight. DDSs are dynamically evolving sources of data changing over time that have the potential to spur real-time action. They are streams of real-time information: a person's mood is captured through a tweet, a restaurant experience is reported in a Yelp review, the presence of an individual at a particular venue is represented by Foursquare's "check-in."

Most critically, people are using mobile devices with increasing frequency for all purposes, creating countless data flows. At the end of 2011, there were six billion mobile subscriptions, equivalent to 87% of the world population,[3] and 12% of these were smartphones. Recent estimates forecast an 18-fold increase in mobile global data traffic by 2016, in the form of video, web and other multimedia content.[4] We believe that it is possible for businesses to take advantage of the real-time nature of these newly generated digital streams.

Our research focuses on the opportunities that arise when companies make use of real-time digital data flows. In this article, we provide a set of frameworks to make sense of DDSs and their potential to foster innovation. Although the field is still new and developing, we provide examples of the opportunities provided by DDSs and describe state-of-the-art uses of digital data streaming.

3   *ICT Facts and Figures*, International Telecommunication Union, 2011.
4   *Cisco Visual Networking Index: Global Mobile Data Traffic Fore-cast Update, 2011-2016*, Cisco Systems, Inc., February 2012.

The Anatomy of a Digital Data Stream

As shown in Figure 1, we have identified three stages of how a DDS comes to be and begins to flow.

Stage 1: Generate

This is the stage at which the digital information is created as an event occurs—whether it be a tweet, a Google search or the GPS position of an object. For simplicity we call this real-time digital representation of events *Digital Data Genesis* (DDG). When a DDG event is not isolated, but rather is part of a series or a stream of DDG events (e.g., sensor readings, readings from smart meters or a Twitter feed), there is an opportunity to generate a DDS. For example, a single click on a hyperlink to access a website is the digital representation of a person's action (a DDG event). Activity on the website is a stream of personal decisions, and is aptly called the clickstream, which is a well-known example of a DDS.

Stage 2: Stream

When the data is available, channeled and transmitted as a continuous flow, we refer to it as a DDS. The streaming stage concerns the manner and format in which the data is made available. The stream is characterized by:

1.   The type of technologies used to create the channel (e.g., application programming

interfaces (APIs), web crawlers, screen scrapers)

2. The nature of the content (including its accuracy and time span)

3. The source of the data being streamed (e.g., public, business, individual or community)

4. The legal status of the data contained in the stream or derived from it (e.g., rights and sensitivity).

Stage 3: Harvest

At this stage, an organization taps into the DDS and extracts some or all of the data being streamed. The harvest stage is described in terms of the technologies adopted to perform the data harvesting (e.g., APIs, XML messaging, web crawlers, screen scrapers).

In isolation, a DDS typically does not contain enough information to enable value creation. As a consequence, the harvest stage includes a process of adding *context* to the data to augment the information thus acquired. The process of adding context may rely on static information or other DDSs.

In a later section, we provide examples of how this context-creation process occurs. The key point is that the process of harvesting a DDS needs to include both the extraction of valuable data from the DDS, and extracting, refreshing and linking of the appropriate elements of the contextual data.

How Firms Create Value with Digital Data Streams

A firm creates value when it provides customers with a product or service they are willing to pay for. In addition to paying for the service or product, customers often have to expend time and effort when selecting one product over another. A firm creates value with a DDS when it leverages the DDS to increase customers' willingness to pay for its product or service or to reduce the opportunity cost of the resources it uses to create existing value propositions.

Below, we describe the five value-creation mechanisms or *value archetypes*, which are essentially "templates" that allow organizations

to create value using DDSs. A value archetype is a basic model for potential value creation with a DDS. Each of the five archetypes represents a class of value-creating initiatives that may employ one or more DDSs. What enables these data streams to contribute to value creation is a *value driver*, which is the unique quality of an activity that is the source of the activity's value contribution. We have identified four value drivers.

Value drivers represent the tangible elements of a DDS that enable a firm to achieve the goals of either increasing a customer's willingness to pay or reducing the opportunity cost of existing resources. DDSs, and the value-creation opportunities they create, surround modern organizations. However, the opportunities are latent in the business environment until DDSs are recognized as tangible resources. A value driver is the mediator between the potential for action and an actual business implementation (i.e., a value archetype). Value can then be created through a mechanism rooted in one or more value drivers.

The five archetypes and four drivers are depicted as a value tree, with the drivers providing the roots for the archetypes (see Figure 2).

Figure 2: The DDS Value Tree

Five Value Archetypes

A value archetype is a basic model for value creation that is based on digital data streaming. Value archetypes are not mutually exclusive, and each organization may create value by using more than one approach simultaneously, as evident in the list of DDS initiatives shown in the Appendix.

1. DDS Generation. A firm may create value by originating the stream of data itself, either knowingly or as a byproduct of other activities. A good example of this value archetype is TripIt,[5] a travel itinerary organization service. Once an itinerary has been created on TripIt, it can be streamed to other partners who can harvest it and create value-added services based on the TripIt platform. Expensewatch.com, for example, integrates a TripIt itinerary automatically to compute expense reports (the Appendix lists further examples of the DDS Generation archetype).

2. DDS Aggregation. With this value archetype, a firm focuses on collecting, aggregating and repurposing a DDS. Socrata (http://www.socrata.com) does this in the U.S. with federal data, acting as a major platform for Open Data[6] and Data.gov initiatives. Federal agencies and organizations generate DDSs, whereas Socrata aggregates them, making them available to the public. Note that the aggregated data made available is created by others.

3. Service. With the Service value archetype, a firm uses one or more DDSs to provide services to consumers or to improve service quality. TripIt takes a step in this direction, and applications like MyCityWay (http://www.mycityway.com/) aggregate hundreds of datasets and DDSs (originating from other entities, such as government sources or the restaurant reservation intermediary OpenTable) to provide consumers with a convenient real-time menu of services available nearby. Similarly, myTaxi (http://www.mytaxi.net) provides reservation and taxi services mainly by collecting the GPS coordinates of both taxis and customers through a mobile application.

4. Efficiency. The Efficiency value archetype uses real-time data streams to optimize internal operations or to track business performance (e.g., waste reduction, response speed). One example is the service made available by Ruter (http://ruter.no/),[7] for the city of Oslo, Norway, and the county of Akershus. Originally known as Trafikanten, this application was created to gather real-time information on the overall state of the city's public transportation system (buses, ferries, trams, metros and trains). Although the intention was to create service value by providing a convenient source of public transport information, the application provided Ruter with real-time status of the city's traffic. This enabled it to optimize the traffic signal priority system, which improved travel times, reduced the number of vehicles and, in general, lowered the costs of transportation. Ruter estimates that the average journey time of buses on some of the most heavily used routes was reduced by up to 20%.

5. Analytics. This value archetype extends beyond Aggregation, with analytics firms processing DDS information to produce analyses or improved visualizations with the objective of enabling better decision making and producing superior insight or knowledge (for example, through dashboards and data mining). Mint (http://www.mint.com/), which is now owned by Intuit, provides an example. In addition to providing its seven million customers with a unified view of their bank, credit and investment accounts, Mint provides them with convenient visual tools to examine the status of their expenses and investments, and with personal recommendations based on their profiles. Mint does this by tapping into the DDSs of the financial institutions with which the customers already have relationships.

Four Value Drivers

Within the framework of the five value archetypes, companies need to focus on and understand the value drivers for each archetype—the enabling factors behind the business opportunity. By identifying value drivers, companies can use DDSs for innovation as they identify new possibilities to create value. In our research, we identified four value drivers.

---

5 Applegate, L. M., Piccoli, G. and Brohman, K. "TripIt: The Traveler's Agent," *Harvard Business School Case 808-059*, October 22, 2008. TripIt is a one-stop travel organization service that allows users to keep track of their travel plans by forwarding their itineraries and confirmation emails to plans@tripit.com. TripIt then integrates all of those plans into a single itinerary that can be accessed from any Internet-connected device.

6 Open Data initiatives are oriented to distribute and make accessible a portion of the data they collect or produce. The basic concept is that by publicizing data, they could be transformed from assets into productive information resources. Several public sector organizations use Socrata's purpose-built cloud, platform and social technologies to deliver citizen access to information.

7 Ruter is the name of the public transport system in Oslo and the surrounding county.

1. Real-time Sensing. This value driver requires the ability to detect the current state of a given entity. Examples are the location of a plane, the speed of a car or the mood of an individual. We consider the real-time sensing value driver to be the *first-order* value driver, because it is the basis of all of the new value-creation opportunities we discuss in this article. For example, the interactivity of Web 2.0 applications and services such as social networks has created new ways for customers to express themselves in real time. Information harvested from Twitter feeds, blogs and Facebook status updates is now used for brand-monitoring purposes. As the interaction with digital systems increases, companies have an unprecedented ability to automatically and continuously sense the world.

2. Real-time Mass Visibility. This *second-order* value driver is based on real-time sensing. It represents the ability to identify the state of multiple entities in real time, contextualized by their relationships. For example, if real-time sensing makes it possible to locate one vehicle, it is possible to acquire visibility of all vehicles on a road, thus enabling traffic congestion to be detected. An example of how real-time mass visibility of vehicles can be used is ruter.no, which enables the provision of added-value services to vehicle operators in the form of real-time traffic information and more efficient routes. Other examples are TomTom (http://www.tomtom.com), which we discuss below, and Inrix (http://www.inrix.com). TomTom, Inrix and ruter.no aggregate information from a multitude of DDS sources, including telecom operators, road sensors and navigation systems.

3. Real-time Experimentation. Another second-order value driver that also relies on real-time sensing is real-time experimentation. It consists of the possibility to fast cycle reliable data generation and gathering. Comparisons of a control sample to other samples (i.e., A/B tests) on web pages for selecting a layout, or the massive experimentation ongoing in major websites, are examples of this value driver. Through real-time sensing, it is possible to experiment and provide immediate feedback on business decisions, from the change of a webpage layout to more complex information. For instance, newBrandAnalytics (http://www.newbrandanalytics.com) provides a service that extracts specific feedback from customers' unstructured mentions of a firm's products or services on social media channels. Firms can then adjust their behaviors in real time, correct any shortcomings and monitor the outcomes. At the same time, they can experiment with different configurations of the service and fine-tune it on the basis of customers' mentions.

4. Real-time Coordination. The third and final second-order value driver is real-time coordination, which is also based on real-time sensing. This driver is the ability to quickly adjust behavior based on feedback about the current state of other entities. Real-time coordination is at the heart of services such as Foursquare Radar, which enables users of the service to coordinate spontaneously with friends by "sensing" their presence in the area.

In addition to the DDS initiatives quoted above using the four value drivers, many other examples are described in the Appendix.

Value drivers are more fluid than value archetypes. Crucially, value drivers change over time, because they are by nature moving targets. Perhaps the most important insight regarding value drivers is that a company does not have to actually own the data generation asset (the DDS). Instead, all the firm needs is access to the DDS and a concept for how to use that DDS to add value. The criteria for success are based on an organization's ability to identify valuable DDSs, gain access to them with the appropriate tools, properly orchestrate the available resources and invest in DDS initiatives.

Examples of DDS Exploitation

As mentioned earlier, TomTom provides an example of opportunities created by the value-creation archetypes and value drivers. Over the past decade, TomTom has been the market leader in personal navigation devices worldwide, achieving a 45% share of sales in Europe and 21% in the U.S.[8] TomTom experienced fast growth since the introduction of its first device in 2004, with revenues soaring from $39 million in 2003 to $1.9 billion in 2007. Its success came from the design of a ready-to-use device that exploited the availability of GPS signals. In 2006, TomTom also introduced TomTom Mobility for real-time traffic information and TomTom WORK for fleet management.

---

8   Garside, J. "Harold Goddijn: TomTom's founder needs his business to turn the corner," *The Guardian*, November 24, 2011.

In 2007, at the height of its global expansion, TomTom introduced HD Traffic, which provides real-time traffic information, and Map Share, which offers personal and community-based maps. This move marked a shift toward a content business model, a trend that continued the following year with the acquisition of Tele Atlas for $2.9 billion. The move to a content model was timely, as sales of TomTom's personal navigation devices began to falter in 2008[9] due to competition from smartphones, automobile manufacturers and a weak economy. For this reason, TomTom now focuses on DDS-based services rather than devices, especially real-time services: *"HD Traffic, our real-time traffic solution, plays an important part in our strategy to expand the services we can offer our customers."*[10]

At the heart of HD Traffic are multiple external data streams from multiple sources that TomTom combines to provide its customers with precise real-time traffic intelligence. These sources include GSM probe data, GPS probe data, incident context data, Traffic Message Channel third-party messages and historic data on average traffic speeds. All these sources are then combined by TomTom to provide better routing compared to competitors and non-interconnected devices.

As of 2012, personal navigation device sales contributed 35% of TomTom's overall sales, so the company is clearly generating revenues from other sources. However, transforming a large company from a device manufacturer and marketer into an information provider is no simple feat. Capturing the potential of DDSs appears to hold the key to TomTom's survival.

We also mentioned TripIt, which is now part of Concur Technologies, in our discussion of value creation. TripIt aggregates a traveler's entire itinerary information not by having access to the global distribution system that has been the backbone of the world's travel system, but rather by tapping into external DDSs. TripIt builds itineraries by monitoring traveler confirmation emails that suppliers send in response to a booking (e.g., airlines, cruise lines, hotels, car rental companies).

With access to the traveler's itinerary, TripIt provides a wealth of information services, making it possible for a customer to download all information about the proposed trip on a smartphone, and even automatically rebook when a flight is delayed enough to miss a connection. When a traveler emails a hotel reservation to TripIt, the firm's proprietary parsing software extracts all information relevant to the itinerary from the message, including the hotel's location. TripIt then adds contextual information to the location by using applications that identify landmarks and points of interest, among other processes. Once it has added this contextual information to the DDS of confirmation emails, TripIt is able to produce a master itinerary with automatically computed useful information such as points of interest, weather forecasts and driving directions.

Note that TripIt creates the contextual information using static data. In other business models, the context is itself dynamically generated from a DDS. This is the case with Foursquare's Radar, which provides a dynamic user-generated database of places to match GPS coordinates of users.

In the same way that a wide range of firms has recognized the value of using GPS data, we anticipate that the potential of digital data streams will no longer be the province of a few early movers, but will soon move into the mainstream. The goal for the firms currently using DDSs is to continue to search for DDS opportunities based on the strengths of their existing assets and capabilities. We believe that TomTom's move from devices to services is an example of this type of evolution.

IT Capabilities Needed to Implement a DDS Strategy

Real-time data streams are low-latency, potentially high-volume flows of digital events. They enable a rapid cycle between the occurrence of an event and a possible (re)action, making them ideal candidates for innovative products and services. The relative simplicity of recording, relaying or analyzing the status of a large number of entities provides substantial opportunities for value-creation. Dealing with real-time DDSs requires a different set of organizational capabilities than dealing with a mountain of "big data." Developing these capabilities is the responsibility of the CIO. Our early research suggests that, to implement a DDS strategy, a firm must have a sound footing in the following four areas:

---

9   Preuschat, A. "TomTom seeks a route back to growth," *The Wall Street Journal Europe*, May 21, 2012.
10 *Annual Report and Accounts 2010*, TomTom NV, 2010.

1. *Dataset:* A DDS strategy hinges on being able to tap into valuable sources. A firm must develop the ability to identify promising internal and external DDSs.

2. *Toolset:* Once a source is identified, the firm must be able to tap into the streaming data. This requires the capability to use appropriate tools to harvest the DDS or multiple DDSs.

3. *Skillset:* A DDS strategy consists of more than simply collecting DDS data. As the five value-creation archetypes show, the firm must develop the competencies to orchestrate the complementary resources necessary to deliver value based on the DDS.

4. *Mindset:* A critical component of innovative initiatives is the willingness to invest and face risks. A hallmark of the successful DDS initiatives we studied was that the mindset of organizational members made them ready to embrace change.

Recommendations for CIOs

To summarize, we offer the following recommendations that we believe CIOs should consider when planning, designing and launching their DDS strategies.

1. Proactively Examine Opportunities to Leverage Existing DDSs

We recommend proactively scanning for DDS-enabled opportunities for value-creation. We believe that CIOs, with their unique blend of technical skills and strategic vision, are ideally placed to identify early on the opportunities for exploiting DDSs. The value archetypes and value drivers described in this article are the blueprints for detecting and envisioning these opportunities.

Once the firm has identified a candidate value archetype, and analyzed the appropriate value drivers, it is necessary to match the recognized potential with available organizational capabilities. As we suggested, these should be articulated in the capacity to identify valuable DDSs, to tap into the information with the appropriate tools, to properly orchestrate the available resources, and to have the willingness to invest in DDS initiatives.

2. Develop the IT Capabilities Needed for the Unique Characteristics of DDSs

The successful implementation of planned DDS initiatives requires an organization to develop the four areas of IT capability described above—dataset, toolset, skillset and mindset. Some organizations, created to exploit the opportunities offered by DDSs, have to create the four capabilities in order to have a product to bring to market (e.g., TripIt). In others, such as TomTom, the development of the appropriate dataset, toolset, skillset and mindset requires the conversion of past practices and assets—which can be even more complex than outright development.

3. Consider Developing a New DDS

The focus of this article is on exploiting existing external sources of data. However, in our research, we also identified many examples of successful innovation through the proactive development of new DDSs. Although this was typically done by startups (e.g., myTaxi), we recommend that CIOs also explore opportunities for generating in-house any data necessary to power valuable DDSs.

A dataset capability is central to this approach, as the value of the DDS cannot be tested until investment has been made to generate it. However, first-mover advantages and barriers to erosion of the advantage could enable those who develop their own DDSs to reap significant rewards. Twitter, Facebook and TripIt are examples of companies that have developed a platform built on top of an internally generated DDS.

4. Be Alert to DDSs your Organization Already Has

While many of the examples of DDS innovators are startups, we are convinced that established firms may have an advantage in this space. Existing firms may well have a dormant DDS potentially waiting to be unleashed. Consider Netflix. After the company introduced its movie streaming service, it had the opportunity to collect and gather additional data on top of simple sales and rentals (for example, in addition to which movie a customer watched, *when* they watched it, *how much* of it they watched). At the same time, Netflix found itself dealing with a new service: *"Streaming has changed not only the way our members interact with the service, but also the type of data available for use in our algorithms.*"[11] Without realizing it, Netflix had created the DDS for its new services.

The pervasive digitization that has occurred in large organizations over the last two decades has made most organizational activities DDG events, and many of them involve streaming. We have identified three steps for taking advantage of this dormant potential:

1. Enable and spur your firm to develop an awareness of the DDSs it has, even though it may not realize it has them. This could be done through a formal inventorying exercise.

2. Identify the value drivers enabled by the DDSs the organization already produces.

3. Match one or more value archetypes to the most promising uncovered value drivers. Once a specific value driver is identified, it is straightforward to develop a business case for the initiative and test whether full development is warranted.

Concluding Comments

In this article, we have explored the value-creation opportunities emerging from digital data streams (DDSs) and provided an initial glimpse of the major managerial implications of this trend. We have also provided recommendations on how to move forward and take advantage of the opportunities provided by DDSs—both external and internally generated streams. In analyzing this new phenomenon, we have attempted to go beyond existing categories and explore new conceptual tools to understand it. Our research has shown that there are significant opportunities for value-creation once organizations have identified digital data streams and treat them as a coherent unit of analysis.

Digital data streams challenge the way organizations have historically extracted value from data. One of the major difficulties for established firms to overcome is that their operations are typically designed for batch processing. But they are now surrounded by constantly evolving streams of real-time data. Harnessing the power of these real-time data flows requires a shift in mindset. We believe that those who can make the shift will find that they can compress the time between the detection of and response to problems or opportunities.

---

11   Amatriain, X. and Basilico, J. "Netflix Recommendations: Beyond the 5 stars," *The Netflix Tech Blog,* http://techblog.netflix.com/2012/04/netflix-recommendations-beyond-5-stars.html.

Appendix: Initiatives, Archetypes and Value Drivers

Initiative	Archetypes	Description	Drivers
Booking.com	DDS Generation	Generates its own real-time stream of room availabilities (199,504 directly contracted hotels). Gathers hotel availabilities through its extranet and allocates them through its website. The generated stream is made available to affiliates, which are rewarded on the basis of an originating fee.	
	Service	Provides its online final customers with a convenient way to benchmark hotel room prices and book online.	
	Efficiency	Improves room occupancy rates for partnering hotels.	Real-time Sensing
		Detects in real-time room availabilities in hotels (and relative prices). Reminds hotels of the number of free rooms to "hurry" the purchase. Provides additional real-time services like last-minute offers.	
		Leverages the website clickstream, for example to hurry purchases by showing the number of visitors currently looking at the same offer.	Real-time Sensing
Evasori.info	DDS Generation	Users can publish online where tax evasion events occur and show them on a map. Users then generate the digital representation of the evasion, and the site streams it. Simple analytics are available, and data can be accessed through a proprietary API.	
		Publishes in real-time tax violations as they occur, including their geographical location.	Real-time Sensing
Groupon	DDS Generation	Generates its own deals DDS by having businesses create and launch featured deals. The DDS is accessible through an API.	
	Service	Aggregates deals and makes them easily accessible thorough web, API and email.	
		Allows the detection of featured deals in a town. Exploits customer location information to geographically filter the deals.	Real-time Sensing
		Provides real-time coordination of users' purchasing behavior. When users find an interesting deal and share it on a social network, coordination emerges through users' connections.	Real-time Coordination
TripIt	DDS Generation	Automatically generates itineraries accessible through an API for third-party use and services development.	
	Service	Organizes and shares customers' trips by creating a master itinerary for access on a smartphone, calendar or anywhere online. TripIt Pro acts like a personal travel assistant that, by aggregating several external DDSs, keeps travelers informed of flight status, alternate flights and more; TripIt for Business is an easier way for companies to organize travel, keep track of who's traveling when and where, and whether travel dollars are being spent wisely.	
		Allows the harvesting of several real-time DDSs to organize the master itinerary, gathering real-time information on flights status, alternate flights, frequent traveler points and eligible flights for fare refunds.	Real-time Sensing
		Provides real-time visibility about travel of a group of individuals who are connected (e.g., co-workers, friends).	Real-time Mass Visibility
Adaramedia	DDS Aggregation	Leverages proprietary data from trusted sources to connect ads with travelers as they surf the Internet. Using exclusive data, reaches premium audiences of travelers and business decision makers.	
		Real-time identification of website visitors provides better advertising targeting, increasing impressions efficiency.	Real-time Sensing

Initiative	Archetypes	Description	Drivers
Factual	*DDS Aggregation*	An open data platform for application developers that leverages large-scale aggregation and community exchange. Makes available datasets on entertainment, health and education, as well as points of interest.	
		Provides access to several real-time DDSs of events or entities' states. As an aggregator, Factual exploits the real-time sensing value driver of other firms or individuals.	*Real-time Sensing*
Influence Explorer	*DDS Aggregation*	Provides an overview of campaign finance, lobbying, earmark, contractor misconduct and federal spending data. Data, accessible through an API, is provided by the Center for Responsive Politics, the National Institute for Money in State Politics, Taxpayers for Common Sense, the Project On Government Oversight, the EPA and USASpending.gov.	
		Allows real-time monitoring of campaign finance, lobbying, earmark funding, contractor misconduct and federal spending data.	*Real-time Sensing*
Infochimps	*DDS Aggregation*	Connects internal data with external data sets accessible through an API. The goal is to streamline access to the world's structured data.	
		Provides access to several real-time DDSs of events and entities' states. As an aggregator, Infochimps exploits the real-time sensing value driver of other firms or individuals.	*Real-time Sensing*
Pachube ("patch-bay")	*DDS Aggregation*	Web-based service that manages real-time data and allows sharing, collaborating and making use of data generated from devices. Acts as a real-time data brokerage platform for the Internet of Things, providing most of its functionality via its API.	
		Allows the sensing of real-time DDSs generated by devices and sensors.	*Real-time Sensing*
Socrata	*DDS Aggregation*	Acts as a major platform in the Open Data and Data.gov initiatives. Federal agencies and organizations generate their own DDSs, whereas Socrata aggregates the DDSs and makes them available to the public.	
		Allows the sensing of several real-time DDSs generated by federal agencies and other organizations. As an aggregator, Socrata exploits the real-time sensing value driver.	*Real-time Sensing*
Spokeo	*DDS Aggregation*	A people search engine that organizes vast quantities of white-pages listings, social information and other people-related data from a large variety of public sources and social networks.	*Real-time Sensing*
		Allows the sensing of several real-time DDSs generated from public sources. As an aggregator, Spokeo exploits the real-time sensing value driver.	
BrandWatch	*Service*	Offers a full range of social media monitoring tools and services, from a simple monitoring project to reports or API integration.	
		Uses real-time sensing of comments about a brand posted on several social websites to gain visibility of the current sentiment relating to a particular product or service.	*Real-time Mass Visibility*
Canal+ Eureka	*Service*	Analyzes 30 million data points a day for the operators of two million internet TV viewers. Tells viewers, for example, that if they liked a particular documentary, they are likely to enjoy another program.	
		Gains mass visibility through real-time sensing of TV viewing, and makes recommendations based on this visibility.	*Real-time Mass Visibility*

Initiative	Archetypes	Description	Drivers
MyCityWay	Service	A real-time app that combines urban reference apps and app platforms to provide a new service to customers.	
		Allows the sensing of several real-time DDSs generated from the status of other entities.	Real-time Sensing
MyTaxy.net	Service	Generates a DDS, which is used to create new convenient services for customers booking taxis.	
	Efficiency	Increases taxi drivers' efficiency by supporting call collection and cab positioning.	
		Detecting a taxi's and customer's positions allows them to be coordinated in real time.	Real-time Coordination
Netflix Recommending system	Service	Creates value through data streams in form of personalization. Uses self-generated and external DDSs, and social networks, to recommend personalized suggested titles based on a household's preferences.	
		Gains mass visibility through real-time sensing of users, and makes recommendations based on this visibility.	Real-time Mass Visibility
TomTom HD Traffic	Service	Provides advanced navigation services and real-time routing and traffic information.	
		The aggregation of data from several real-time DDSs (GSM probe data, GPS probe data, incident context data, TMC third-party messages) provides visibility on traffic congestion.	Real-time Mass Visibility
ruter.no (Trafikanten)	Service	Gathers real-time information on the overall situation of Oslo's public transportation and repurposes it in a convenient web and mobile application to make the lives of commuters and citizens easier.	
	Efficiency	Optimizing traffic signal priority system based on real-time information about the city's traffic improves travel times, reduces the number of vehicles and lowers the costs of transportation.	
		Real-time sensing of the DDSs from public transport operators provides visibility of the traffic levels throughout the whole system.	Real-time Mass Visibility
American Car Rental	Efficiency	Automatically charges customers exceeding speed limits by monitoring customers' driving speeds in real time.	
		Allows real-time monitoring of a car's speed (and other telemetry information).	Real-time Sensing
Autostrade Tutor	Efficiency	The police can tap in the camera system used to manage Italian highways to automatically fine speeding drivers.	
		Allows the average speed of each vehicle passing under consecutive "Tutor" enabled gates to be known in real time.	Real-time Sensing
Google Trends	Analytics	Analyzes a portion of Google web searches to compute how many searches an individual has done for particular terms relative to the total number of searches done on Google over time.	
		Allows real-time sensing of Google users' web searches.	Real-time Sensing
Mint	Analytics	Provides a unified view on bank, credit and investment accounts by connecting to the DDSs of the originating financial institutions. Value is created as customers get a unified view of their finances, convenient visual tools to examine the status of their expenses and investments, and personal recommendations based on a specific profile.	
		Gains mass visibility of current customers through real-time sensing of DDSs, and makes recommendations based on this visibility.	Real-time Mass Visibility

Initiative	Archetypes	Description	Drivers
newBrand-Analytics	*Analytics*	Extracts specific feedbacks from customers' unstructured mentions of brands on social media channels. Firms can then adjust their behaviors in real time on the basis of customers' mentions.	
		Allows fast cycling from customers' feedback to action, enabling firms to fine-tune their offerings on the basis of customers' mentions.	*Real-time Experimentation*
Opower	*Analytics*	Compares the data from a customer's smart meter (anonymously) with that of neighbors, using graphics, bar charts and SMS alerts. Customers are presented with information about behaviors and how to reduce their power consumption.	
		Leverages the grid's smart metering capabilities to provide neighborhood consumption visibility and thus promotes energy-efficient behaviors.	*Real-time Mass Visibility*
TaKaDu	*Analytics*	A software-as-a-service solution for monitoring water distribution networks. Provides the utility with real-time control over network events, using state-of-the-art statistical and mathematical algorithms. Detects, alerts and provides real-time insight on leaks, bursts, zone breaches and other network inefficiencies.	
		Allows real-time sensing of the water distribution network	*Real-time Sensing*

About the Authors

Gabriele Piccoli

Gabriele Piccoli (gabriele.piccoli@gmail.com) is Associate Professor of Information Systems at the University of Pavia (Italy). Previously, he held positions as Full Professor at the Grenoble Ecole de Management (France) and Associate Professor at Cornell University and the University of Sassari. His expertise is in strategic information systems and the use of information systems to enable customer service. He is a former Associate Editor of *MIS Quarterly,* and his research has appeared in *Harvard Business Review, MIS Quarterly Executive, Communications of the ACM, MIS Quarterly* and *Decision Sciences Journal,* as well as other academic and applied journals.

Federico Pigni

Federico Pigni (federico.pigni@grenoble-em.com) is Assistant Professor in Information Systems in the Management of Technology and Strategy department at the Grenoble Ecole de Management in France. He was Senior Researcher at Carlo Cattaneo University's Lab#ID RFID laboratory and held a post-doctorate position at France Télécom R&D—Pole Service Sciences (France). He has participated in various research projects funded by Italian, regional and EU agencies, private industry and government partners. He teaches in the area of information systems and has a research interest in the strategic application of information systems in interorganizational contexts and the use of innovative IT to deliver customer service.

Maximizing Value from Business Analytics

CIOs need to maximize the value from the significant investment their organizations make in business analytics (BA) initiatives. We explore two themes for maximizing BA value—speed to insight and pervasive use, and present a BA case study at GUESS? INC., a fashion retailer. We provide recommendations for how IT leaders can maximize value from their BA investments.[1]

Barbara H. Wixom
MIT Sloan School of
Management (U.S.)

Bruce Yen
GUESS?, Inc. (U.S.)

Michael Relich
GUESS?, Inc. (U.S.)

Enterprise Business Analytics Capabilities

Companies increasingly deliver value through business analytics (BA), which includes the people, processes and technologies that turn data into the insights that drive business decisions and actions.[2] As Figure 1 illustrates, organizations with enterprise BA capabilities establish a sound foundation of high-quality, usable and integrated data. This data is delivered to business users via a diverse portfolio of business analytics tools, including query, reporting and advanced analytics software. Business users identify insights from the data, make decisions and solve important business problems, thereby triggering actions that generate a wide range of tangible and intangible business value. The data provided through BA is also known as business intelligence (BI). Over time, organizations manage and evolve their BA capabilities through IT and data governance mechanisms.

The Enterprise Business Analytics Capabilities Model[3], shown in Figure 1 is a lens through which we can examine how companies are building, managing and changing their business analytics capabilities. From an analysis of emerging practices recently reported by almost two

1 This article is based on research sponsored by the Advanced Practices Council of SIM.
2 Eckerson, W. *Secrets of Analytical Leaders: Insights from Information Insiders*, Technics Publications, 2012.
3 For more information about the Enterprise Business Analytics Capability Model, see Wixom, B., Watson, H. J. and Werner, T. "Developing an Enterprise Business Intelligence Capability: The Norfolk Southern Journey," *MIS Quarterly Executive* (10:2), 2011, pp. 61-71.

Figure 1: Enterprise Business Analytics Capabilities Model[4]

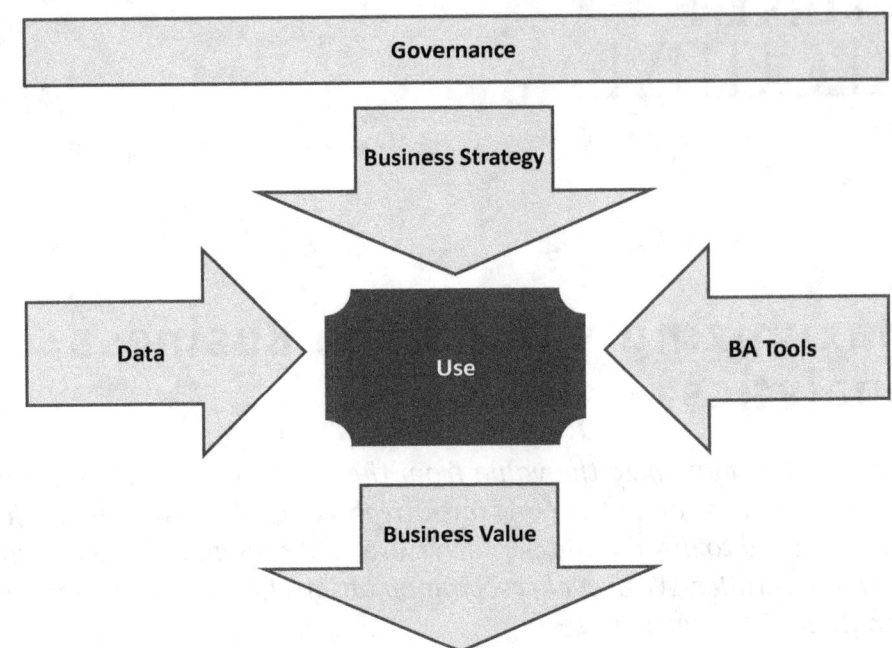

dozen companies, two themes emerged that characterize how companies are attempting to maximize business value from their enterprise BA capabilities: speed to insight and pervasive use. In the following sections, we describe these two themes and some contemporary practices that facilitate them.[5]

Speed to Insight

Speed to insight is concerned with how expeditiously organizations transform raw data into usable information. Practices that facilitate speed to insight can be categorized as automation, business requirements and reuse (see Figure 2 and Appendix 1).

Some practices, such as data standards and metadata, help organizations *automate* data on-boarding,[6] integration and quality processes. The more automated these processes, the faster data can be physically transformed into usable information. For example, a healthcare company initially estimated a project to on-board more

than 30 new data sources into its business analytics environment would take two months. By shifting to configurable, metadata-driven on-boarding processes, the project was accomplished in five business days. And an IT company cut the estimate for a data integration project by 30% after implementing automated data mapping.

Other practices, such as agile development methods, sandbox environments[7] and co-locating developers with business users, enable development teams to more rapidly identify *business requirements* for data and then translate those requirements into business analytics products and services. Until a few years ago, agile development was rarely applied to business analytics projects, but that has changed as companies have seen the positive impact of agile development on delivery schedules. For example, an insurance company moved to an agile development process for all BA projects, adopting techniques like paired programming, story walls and test-driven development. The company realized a four-fold increase in analytics usage over two years, attributed to increased development productivity (i.e., increased BA delivery).

---

4   Adapted from Wixom et al., op. cit., 2011.

5   The primary data source for identifying the two themes was the applications submitted in 2011 and 2012 by 23 companies for the emerging practices category of The Data Warehousing Institute's (TDWI) annual business analytics best practices competition. A summary table with the specific practices reported by each company is shown in Appendix 1.

6   On-boarding is the process of incorporating new data sources into a company's data infrastructure.

7   Sandboxes are technologies outside of an organization's core systems that, in this context, are used by analytics professionals to develop ideas for new applications.

Figure 2: Drivers of Speed to Insight

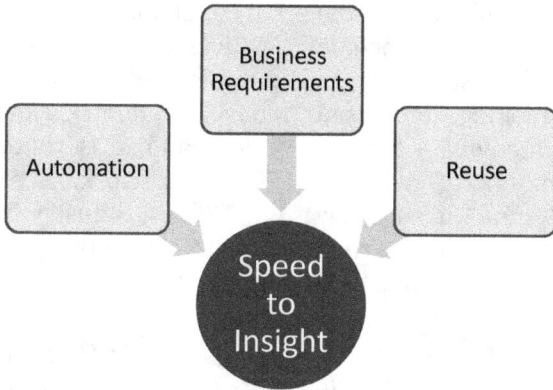

Figure 3: Drivers of Pervasive Use

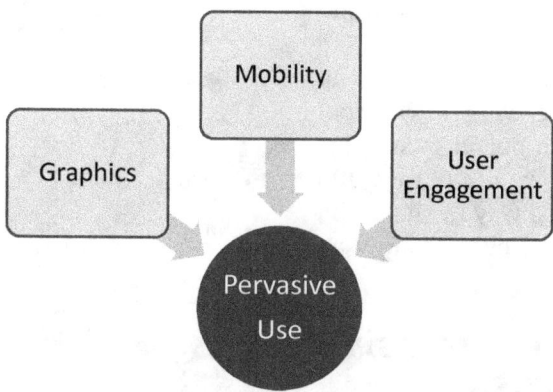

Finally, companies that invest in *reuse*, which includes such practices as data services, design catalogs and parameterized reporting, can get information into the hands of business users more quickly. A healthcare company reduced data on-boarding from 1,700 hours to a few hours by using a data-as-a-service approach. The company estimates that use of data services reduced developer time by 25% and data store redundancy by 75%. Another healthcare company uses a catalog of best practice approaches for designing dashboards to meet various objectives. Use of the catalog shortened the company's delivery time for dashboards from 12 to five days.

Pervasive Use

On average, 25% of an organization's employees use some form of business analytics to do their jobs.[8] Companies can increase this

percentage by adopting practices that encourage more pervasive use of business analytics across the enterprise, such as graphics, mobility and user engagement (see Figure 3 and Appendix 2).

Visually appealing software interfaces using *graphics* encourage pervasive use because the adage "a picture is worth a thousand words" holds true in business analytics. Users react positively to appropriate uses of maps, colorful dashboard displays and advanced visualization approaches. One retailer found that adding photographs of products to reports positively impacted adoption and analyst productivity.

A second driver of pervasive use is *mobility*— delivering business analytics via mobile devices such as cell phones and iPads. Companies reported particularly high adoption success and user enthusiasm with iPad-based BA deployments (as described in the GUESS case below). In general, the key benefits of mobility are portability and ease of access to business analytics. For example, a healthcare company that

---

8   This figure is cited by Cindi Howson in many of her webinars, which can be found at www.biscorecard.com/businessintelligencewebinars.asp.

transitioned from web-based BA to mobile BA attributed part of the benefits it gained from the iPad's "cool factor." Benefits of mobile BA include more frequent analyses, instant decision making, more consultative decision making, increased self-service and increased productivity. A biotech company estimates that members of its sales force save between 30 and 90 minutes each day through its mobile BA application, and that time saving translates into $4 million annual savings or increased productivity.

Finally, companies can deepen BA usage through practices that promote *user engagement*. This broad category includes self-service approaches (e.g., report wizards), gamification[9] (e.g., incentivizing use) and collaboration techniques (e.g., rating, discussion and sharing portals)—all of which draw users into BA and engages them in an interactive way. An Internet company set up a collaboration portal to support its anticipated BA growth from 1,000 to 4,000 users. The portal offered users a way to share and rate analyses, discuss ideas and even "follow an analyst."

Speed to Insight and Pervasive Use at GUESS

For more than 30 years, GUESS?, INC. (referred to as GUESS) has been designing, marketing, distributing and licensing collections of contemporary apparel and accessories for men, women and children. A $2.5 billion company, GUESS competes globally in the fashion retail industry, operating in 87 countries.

GUESS operates using a variety of business models that vary geographically. The U.S. is predominantly a retail business, with sales to consumers accounting for 85% of business and 15% going to wholesale (e.g., department stores). Europe is about 75% wholesale. Asia is a mixture of retail and wholesale. In Central and South Americas, GUESS engages in partnership arrangements. Although these diverse business models require localized business processes, GUESS centrally controls its brand and delivers a consistent customer experience across its distribution channels.

GUESS succeeds by placing the right apparel in the right store at the right time to appeal to its fashion-savvy shoppers. To do this well,

the company needs to be good at fashion and at distribution. To accomplish the former, GUESS employs designers who identify fashion trends and create appealing styles. A staff of buyers, planners and distributors ensure that merchandise is routed appropriately across the GUESS network.

The BA industry has recognized that GMobile, GUESS's business analytics iPad initiative, is an innovative and game-changing BA application.

Table 1: Practices to Drive Speed to Insight and Pervasive Use at GUESS

Speed to Insight	Pervasive Use
• Data Standards	• Photographs
• Agile methods	• Advanced Visualization
• Co-location	• Dashboards
• Shadowing	• Graphic designers
• Templates	• iPads
	• Collaboration

GMobile uses several practices that facilitate speed to insight and pervasive use of business analytics (see Table 1).

The Origin of GMobile

GUESS's CIO and his BA director initiated the GMobile project while attending a BA vendor conference. As the conference keynote speaker described the potential value of the iPad for BA delivery, the CIO realized that the iPad's portability, graphical nature and trendiness could be a good fit for his highly visual and creative business users. During the talk, he texted GUESS's procurement department to order several iPads for his team so they could begin exploring the iPad's potential when he returned.

"Twenty minutes into the conference presentation, I was convinced that this was the perfect device for my merchants to consume data. They are always running around with 5- or 6-inch binders filled with hundreds of pages of 8-point text with every little metric—and they still seem to be missing key pieces of information. If we

---

9   Gamification is the integration of game mechanics or game dynamics into an information system.

built them a buyer's workbench, it would be the perfect application." CIO, GUESS

At the time, GUESS had significant BA capabilities in place. For ten years, the company has had a data warehouse that supported ad hoc analyses, BlackBerry reports and web-based dashboards. The BA director, who previously worked for a BA vendor, not only had deep technical expertise, but had a strong working relationship with GUESS's business users. However, the CIO and BA director believed that delivering BA via the iPad could potentially be a game-changer.

Developing GMobile

As a first step, the BA director asked his team to download highly rated iPad apps and identify what made them so popular. The team not only explored media and productivity apps, but also games. For example, a vegetable chopping game showed the importance of color, ease of use and fun. The team found that exploring and discussing apps helped with its understanding of app workflow, the way in which data could and should be delivered (i.e., how much and when), communicating instructions and the effective use of graphics.

The BA director then began shadowing his users "because the scenarios for which we were designing were so different from previous scenarios." The vision for GMobile was a buyer's workbench that would replace binders of reports and support core work for the GUESS knowledge workers responsible for product distribution. The BA director visited stores with users. *"We sat in their meetings. We had never asked them to open up that much to us before, but I think they realized that it would be very cool to have a tool that could help them at those meetings or when they were on the road."*

The BA team engaged a graphic designer to help develop GMobile. *"We wanted the graphic designer to polish the app and make it look really good so that people would be drawn in. We wanted our users to wake up Saturday morning, read the newspaper and look at sales on their iPad app."* The designer helped to implement a visually appealing app that incorporated a Hollywood theme with related graphics and colors. The app also included product photos and geospatial mash-ups.

"We didn't want to create a series of dashboards. We wanted more of a multi-dimensional, interactive workflow where a user can tap and quickly get to more insight and more detail. And you can go back to your beginning point easily. Think of it as one of those old 'Choose Your Own Adventure' books that we read as kids. Flip to page 73. You open the door on the left. Page 14, you open the door on the right. And each choice leads to a different path through the book." BA director, GUESS

The type of information delivered via the iPad evolved over time as users asked for more and different functionality. For example, the app initially provided best-sellers by stores. Over time, users asked for store best-sellers categorized by additional dimensions, such as style and color.

GMobile Data Foundations

GUESS's varying local business models require localized point-of-sale systems and regional ERP systems. The company achieves data standardization through a centralized Product Lifecycle Management (PLM) system that serves as the GUESS system of record. All style information—whether about a fabric, trim or garment—is created in the PLM system and then pushed into the ERP systems to achieve consistency with local execution. Additionally, the three regional data warehouses (for Asia, Europe and the U.S.) use the same data model with common attribute definitions. This ensures that regional reporting is consistent and that data can be integrated into a single global view.

"We built our own data model. I have a core team that has worked for me on average for 15 to 17 years. A lot of these guys have worked for me at four different retailers. They have a lot of retail experience. Since we've looked at many industry data models, we are familiar with what is out there— and know what other retailers have used. We created what we consider the best of breed." CIO, GUESS

The regional data warehouses, running on traditional data warehousing technology, had met GUESS's needs for the past decade, but the IT team found that GMobile generates different technical requirements. iPad users expect

Figure 4: GMobile Screen Shot

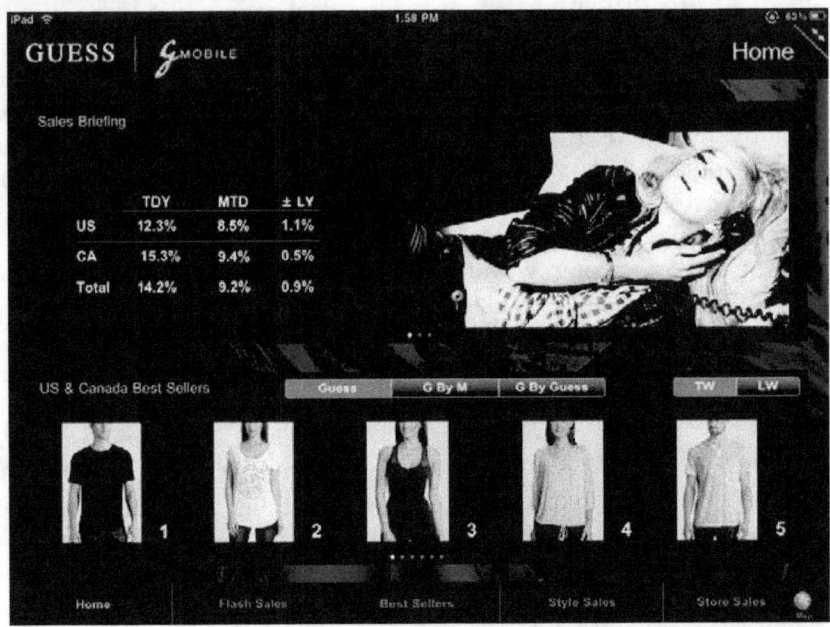

extremely fast response times, yet the iPad does not have the technical capacity to support techniques such as caching. This means that the data warehouse has to close the performance gap.

To improve performance, the BA team ported the North American data warehouse to a column-oriented data warehouse appliance. While the previous data warehouse took 20 to 30 minutes to run complex analyses, such as product affinity or market basket, the column-oriented appliance performed the same analyses in seconds.

> *"We named the new data warehouse 'the Maserati.' I told our users 'We were driving around in a Volkswagen beetle—but now we are running around in a Maserati. It is light years ahead in terms of speed.' ... It's been a huge enabler because GMobile requires that we serve up queries quickly."* BA director, GUESS

GMobile Business Analytics Tools

Prior to GMobile, GUESS's business analytics software portfolio had evolved from ad hoc queries to reports delivered via BlackBerries to web-based dashboards. Throughout this evolution, users became more comfortable and familiar with using reports and dashboards—and, in general, with using data for decision making.

The GUESS IT group believes that the iPad had potential to be a game-changing method for delivering business analytics because it combines the best of all worlds—the portability of a mobile phone, the functionality and screen size of a laptop, and the rich media, interactivity and appeal of a current "hip" technology. The latter proved particularly true. The IT group suspected that some users signed up for GMobile simply to receive an iPad. This did not concern the team because, over time, users with iPads ultimately would become highly engaged BA users.

The iPad's rich media support is important for GMobile. The app incorporates a variety of charts, graphs and maps that depict best-sellers and store sales information. The development team discovered that its decision to include product photographs into the app excited its business users (Figure 4 shows a GMobile screen shot).

> *"It was a breakthrough in realizing that visual analytics didn't mean a geospatial tool or a lot of time creating charts and graphs. There was a missing piece of analytics that we could bring just by doing something as simple as adding a picture instead of listing styles so that the users could visually see what's happening."* BA director, GUESS

Table 2: GMobile Business Value

Transactional	Informational	Strategic
• Less paper • Time savings • Fewer meetings • Reduced headcount • Faster cycle time • Convenience	• Factual decisions • Real-time decisions • Single version of the truth • Business pattern discovery • More collaboration	• Speed to market • Improved business understanding • Reputation

The development team leveraged two important roles when creating GMobile. The first was a graphic designer, who worked with the IT group to make the screens attractive and consumable. The designer incorporated a "fun, Hollywood theme" into GMobile to create an appealing user experience. Additionally, the designer focused on making the app easy to use by supporting an intuitive workflow and ensuring that users could always find their way back to an earlier screen or to the app home page. This was helpful for the less tech-savvy users, who found the iPads highly approachable.

The second role was an app developer. This person ensured that GMobile leveraged the nuances of the iPad, such as swipes and gestures, within the interface design. The app developer also ensured that GMobile did not simply replicate previous dashboard reports that were developed for the web; in fact, the GMobile app delivers data that was previously reported through 12 different dashboard applications. The iPad supports a more interactive, versatile way to deliver data, which supports a wide range of user work styles and work flows.

> "Different people work in different ways— and they like to see information in different ways. Through the app, you can manipulate and view data however you need it." Director of Mexico & Latin America Support, GUESS

One drawback of deploying BA through the iPad was that the device technology was still fairly new when development began, and many bugs had to be fixed. The team needed to develop workarounds and seek BA vendor assistance to solve issues with memory management, security support and networking. *"We basically had a lifeline to them,"* explained the BA director. The team also had to address "bring your own device"

(BYOD) issues for users who wanted to put GMobile on their personal iPads.

Generating Value with GMobile

The GMobile app generates a wide variety of tangible and intangible business value for the company that can be categorized as transactional, informational and strategic (see Table 2).

Transactional Benefits. GUESS is gaining several productivity improvements from GMobile that result in bottom-line cost savings. As anticipated, the iPad devices have replaced reams of paper reports, reducing paper costs and increasing eco-friendliness.

> "I like that I can carry the iPad and not carry an inch of paper as I did in the past. When information is not in GMobile, I create PDFs and have that available through a PDF app. When the visual team goes out to stores and creates actual windows with product, they take pictures and send them to us. The iPad is a piece of equipment with a lot of information in it." Director of Planning, Retail, GUESS

Since the GMobile app was created to answer many more questions than its predecessor applications, users spend less time finding answers and fewer analysts are needed to prepare reports.

> "Four years ago, we had about 12 planners, and now we are down to seven because we are doing less reporting. And the reporting is more cohesive." Director of Planning, Retail, GUESS

People at GUESS no longer need as many meetings as previously to "get people on the same page." Traditionally, GUESS held a weekly meeting for 40 representatives across the company to discuss best-selling items. Now that meeting is bi-monthly because the GMobile app communicates best-seller information so effectively.

Informational Benefits. As business users adopted GMobile, the IT group observed that the nature of some work began to change, particularly in the way users collaborated and communicated with each other. Since the iPad became most users' primary information repository, they incorporated other kinds of reporting, note taking and even photographs into their decision-making processes. Users now take photos using the iPad to capture store layouts, window designs and even competitor marketing efforts. These photos are then incorporated into future decision processes or referenced in meetings with others. Overall, GMobile provides users with more and better information, leading to improved, more fact-based decisions.

Strategic Benefits. BA also delivers strategic benefits to GUESS. One important benefit is that users develop a deeper understanding of the business. GMobile users cite numerous examples where this understanding has resulted in stronger business performance. One merchandiser used the app to understand how the launch of a new product performed in its very early stages within North America and how that performance translated into South American market performance. This understanding led to better purchasing and distribution decisions, and, ultimately, more sales of higher profitability items in her region. Another user applied BA to identify size profiles for stores, discovering that some stores tend to have customers who purchase smaller sizes and other stores tend to sell a greater number of large-sized clothing.

"Once you incorporate size profiles into your decision processes, you actually increase your sales in every store by some amount, because now all these people who are extra-smalls are not walking [out the store], as opposed to other stores where people who are larges and extra-larges are walking because we didn't give them enough. You increase your business and you reduce your markdowns because you no longer have extra units of some items sitting in a store." Director of Factory Planning, GUESS

GMobile also generates intangible strategic benefits. Use of the iPad app communicates the perception that GUESS operates in a leading-edge and "hip" manner, which resonates well with its many partners around the globe. Additionally, the adoption and popularity of this iPad initiative fosters innovation internally at GUESS, prompting other iPad-related projects elsewhere in the company.

Summary of GUESS's Enterprise Business Analytics Capabilities

Figure 5 summarizes GUESS's key enterprise BA capabilities. The company achieves standardized and high-quality data from its PLM system and enterprise data model, and delivers data for analytics via a column-oriented data warehouse appliance. It also includes photos as a data source to facilitate consumable reporting.

The company offers a wide array of BA tools (e.g., ad hoc queries, BlackBerry reports, web-based dashboards and GMobile) to its executives, designers and merchandisers across the globe. The IT team uses agile development methods and user shadowing to identify business requirements, and leverages a graphic designer, app developer and strong vendor relationships to deliver leading-edge applications that maximize the device used for delivering BA to users.

Business users adopt BA because the tools are easy to use and useful to their work processes; the tools also facilitate collaborative decision-making processes. The GMobile iPad app further engages users by offering an enjoyable experience. The business-savvy IT team and strong IT/business relationship keep the BA efforts aligned with real business needs, resulting in transactional, informational and strategic benefits for GUESS.

Recommendations for Maximizing Value from BA Investments

The following five recommendations—two concerned with speed to insight and three with pervasive use—will help IT leaders maximize the value from their BA investments.

Figure 5: GUESS's Enterprise BA

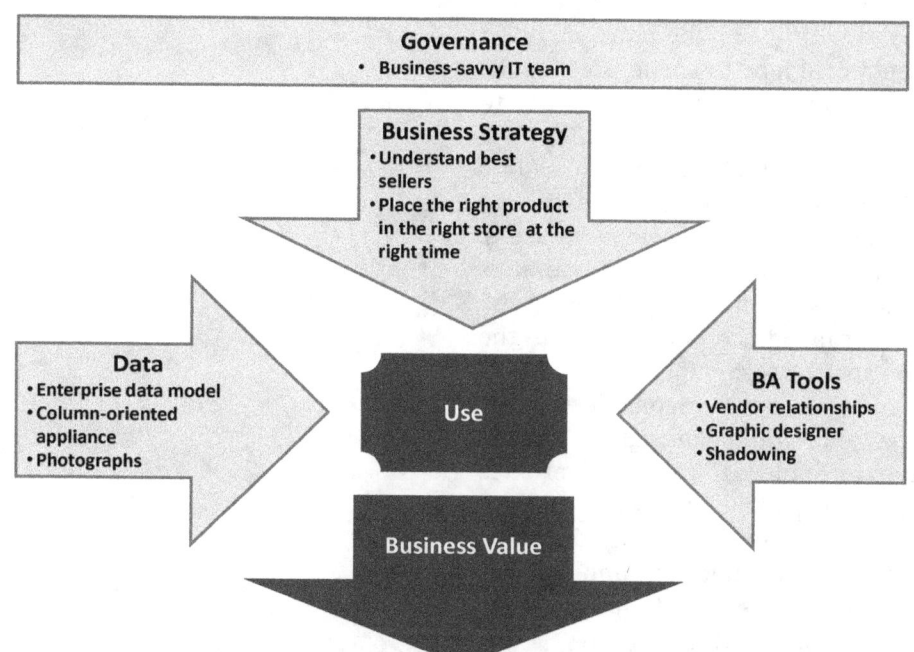

Recommendations to Drive Speed to Insight

1. Create an Optimized Ecosystem of Advanced and Traditional Data Technologies. Organizations should incorporate a variety of database technologies into data architectures as BA techniques evolve and new sources of data become available. There is no longer a "one-size fits all" technology for BA. Thus, after identifying unique data processing needs, IT leaders should invest in a set of technologies to address those needs. The future data architecture will be an ecosystem of technologies that can concurrently process unstructured data, streams of real-time data and large volumes of historical events. Speed to insight will correlate with the IT group's ability to match processing needs with processing capabilities.

2. Develop Data Standards, Even if it Means Creating a Standards Layer on Top of Diverse Systems. Standardizing data at an enterprise level will only get harder as data architectures increasingly evolve to the federated, ecosystem approach. A federated architecture, however, does not reduce the need for data standards. In fact, a data standards layer becomes even more important for automating data integration and data-quality processes to achieve fast speed to insight. If standardized data sources are not possible, consider enterprise data models, an enterprise platform system (e.g., PLM at GUESS) or a master data management initiative.

Recommendations to Drive Pervasive Use

3. Invest in Business-savvy IT Staff. Business-savvy IT professionals are particularly important in BA for two reasons. First, they ensure that business requirements are met; research shows that meeting business requirements drives usage. Second, business-savvy IT professionals ensure that the business requirements correctly address the company's real business needs. Some IT groups are fortunate to have staff with long organizational tenures and deep business knowledge. Those that do not should consider investing in rotation or training programs to develop business-savvy abilities.

4. Encourage User-intensive Development Practices. Even the most business-savvy IT professionals may neither understand exactly what business users do each day nor how and when they make decisions. Practices that help developers understand user work styles and behavior, such as shadowing, agility and co-location, can improve BA development outcomes. This is likely a key reason behind the sharp rise in agile development methods for BA projects over the past few years. The more time that developers spend with users, the more nuanced their understanding of how to shape BA to make tools and applications more useful and easier to use.

5. Exploit the "in" Technology. IT leaders should embrace the new technologies that come to market as a way to excite and engage BA users. Since each device has both unique capabilities and constraints, consider hiring specialists who can exploit the capabilities and mitigate the constraints of a specific technology. For example, to leverage an iPad device, consider hiring app developers and graphic designers skilled in shaping the technology's visual experience to incorporate geospatial awareness and/or photographic images in workflows. At the same time, the iPad's technical limitations and security risks need to be addressed with appropriate BYOD policies and other controls.

Concluding Comments

The Enterprise Business Analytics Capabilities Model offers an approach for articulating key practices that build and shape BA capabilities. Our research suggests that once BA capabilities are established, business value is maximized by using practices that drive speed to insight and by making BA usage pervasive across the enterprise.

The benefits from BA will be both tangible and intangible, ranging from very tangible productivity improvements (such as less paper-reporting and time spent in report preparation) to intangible benefits (such as improved company reputation and deeper strategic business understanding). All of these benefits are important for maximizing BA value.

The experience of GUESS with GMobile shows what can be achieved. Having established enterprise BA capabilities over the past decade, GUESS is now focused on reducing the time it takes to transform data into usable information (i.e., speed to insight) and deepening the usage of BA across the enterprise (i.e., pervasive use). Further, as BA capabilities evolve to incorporate new trends, such as big data sources and cloud-based architectures, CIOs should monitor how these trends can be leveraged specifically to drive speed to insight and pervasive use to maximize value from business analytics.

Appendix 1: Practices that Facilitate Speed to Insight[10]

Company	Project	Speed to Insight Practices
Healthcare	• On-boarding and standardizing new data sources; ongoing data quality control	• Configurable, metadata-driven platform • Ability to self-configure field-level quality control levels • Automated business rules repository
Automotive	• Agile, iterative development for data warehousing	• Automated testing to enable real-time feedback of development changes
Transportation	• Data validation and certification process for on-boarding and sharing data	• Data standards • Business rules engine to manage industry-level data sharing
Financial Services	• Daily data processing performance improvement	• In-database processing • Parallelism
Aerospace	• BI report development	• Lean framework adapted for agile BI • BI competency center • Dedicated/co-located teams
Healthcare	• Enterprise data integration architecture, which supports metadata management and service-oriented architecture	• Model-driven, wizard-based data services • Data standards • Metadata
IT	• Data mapping for integration processes	• Automated data-mapping process • Integration center of excellence
Insurance	• BI reporting	• Agile development process (e.g., co-location, story cards and four-week roll-outs)
Financial Services	• Data integration and delivery	• Virtualization architecture • Data standards • Metadata • Data services
Energy	• "Live" operational dashboards	• Complex event-processing technology • Business activity monitoring • Visualization
Internet	• Data warehousing and reporting on Internet data	• Hadoop[11] solution to process large volumes of unstructured, real-time Internet data
Financial Services	• Compliance reporting	• Metadata-driven reporting architecture that automatically adjusts to complex changes to reporting requirements
Healthcare	• Dashboard reporting methodology	• Dashboard delivery process for highly visual, standardized reporting • Center of excellence • Dashboard design catalog, which promotes visualization best practices
Insurance	• BI reporting	• Agile methodology • High business user involvement
Pharmaceutical	• Master data management	• Master data management platform

---

10 This table lists the practices from companies that applied to The Data Warehousing Institute's 2011 and 2012 best practices competition in the emerging trends category.

11 A software framework that supports data-intensive distributed applications under a free license.

Appendix 2: Practices that Facilitate Pervasive Use[12]

Company	Project	Pervasive Use Practices
Aerospace	• BI reporting	• Graphical delivery • Dashboards • Prompt-based reports • MDX[13] functions
Insurance	• Campaign management	• Automated model scoring • Excel interface • Sandbox environment
Call Center	• BI reporting	• Advanced visualization • Dashboards • Mobile delivery • Collaboration supported by having users rate, comment on and discuss dashboard content
Healthcare	• BI reporting	• Mobile delivery • Advanced visualization • Geographic-specific reporting based on location awareness
Financial Services	• BI reporting	• Parameterized reports • Key performance indicator wizard to generate custom dashboards • Advanced visualization • Gamification framework that wraps a game layer around operational reporting
Restaurant	• BI reporting	• Self-service • Data standards
Biotech	• BI reporting	• Mobile delivery • iPads • Mobility center of excellence • Video training clips
Internet	• BI reporting	• Self-service • On-line portal • Knowledge management, allowing user to share, discuss and rate analytics practices

12   This table lists the practices from companies that applied to The Data Warehousing Institute's 2011 and 2012 best practices competition in the emerging trends category.

13   MultiDimensionalEXpressions—a multidimensional query language.

About the Authors

Barbara H. Wixom

Barbara Wixom (bwixom@mit.edu) is a Principal Research Scientist at the MIT Sloan School of Management's Center for Information Systems Research (CISR). Her areas of expertise include how firms build and deliver business value from enterprise data capabilities. Prior to joining CISR, she was an associate professor at the University of Virginia's McIntire School of Commerce, teaching courses in data management, business intelligence and IT strategy at undergraduate, graduate and executive education levels. She has published in journals such as *Information Systems Research*, *MIS Quarterly*, *MIS Quarterly Executive* and *Journal of Management Information Systems*, and has presented her work at national and international conferences.

Bruce Yen

Bruce Yen (bruceye@guess.com) is Director of Business Intelligence at GUESS?, INC. and leads the business intelligence and data warehousing initiatives. He specializes in creating a cohesive data-dissemination strategy that brings actionable data to diverse user communities and business needs. He has over 13 years of data warehousing and business intelligence experience. Yen is a recognized leader in business intelligence and has received industry recognition as an innovative and cutting-edge information manager for both dashboard and mobile application design and implementation. Earlier experiences include consulting for MicroStrategy and managing the data warehouse for the North American Bottled Water division of Group Danone.

Michael Relich

Michael Relich (mrelich@guess.com) is Executive Vice President and Chief Information Officer at GUESS?, INC. Prior to joining GUESS, he served as CIO and Senior Vice President of MIS and E-Commerce at Wet Seal, Inc., a specialty apparel retailer, and as Senior Vice President, Engineering at Freeborders, Inc., a Product Lifecycle Management (PLM) solutions provider. Relich has also held senior-level IT positions with retailers HomeBase Inc., where he served as Assistant Vice President of MIS, and Broadway Stores Inc., where he served as Director of Merchandise Systems.

MIS QUARTERLY EXECUTIVE

The Evolution of Information Governance at Intel

This article describes the decade-long evolution of information governance at Intel against a background of rapid increases in data volumes. Intel's initial governance model sought to contain risk by restricting access to key information resources. The model evolved to a Protect-to-Enable approach that balanced the need to protect data with the need to make data more accessible and available for decision making. The information governance lessons learned from Intel's experience can be applied by other organizations.[1,2]

Paul P. Tallon
Loyola University (U.S.)

James E. Short
University of California,
San Diego (U.S.)

Malcolm W. Harkins
Intel Corporation (U.S.)

Big Data and the Need for Information Governance

In recent years, the volumes of data captured and retained by organizations have grown exponentially. In some industries, principally pharmaceuticals, healthcare and energy, the volume of retained data is doubling each year.[3] The huge volume and variety of data now being stored is known as *big data*. How to govern access to and use of big data is now a critical concern for CIOs as they face a rapidly expanding flood of new data from sources such as RFID, web transactions and social media.

Three factors underlie the rise of big data. First, access to better, faster and cheaper storage has made it easier for organizations to capture and retain larger amounts of data for longer periods of time. This level of price-performance improvement has increasingly led users to believe that storage is *free*, so they are not motivated to delete unimportant data. Second, firms are seduced by the hype surrounding data analytics and the possibility of uncovering important insights through data mining. This has led users to retain data for long periods even if its analytical value is negligible in the short term. Third, regulations often specify that data be retained for specified periods even if the data has ceased to be useful for decision-making

KELLEY SCHOOL
OF BUSINESS
INDIANA UNIVERSITY

1   Cynthia Beath, Jeanne Ross and Barbara Wixom are the accepting senior editors for this article.
2   The authors thank the editors for their thoughtful comments and suggestions and Terry Yoshii (Intel IT) for his assistance in coordinating interviews with Intel IT personnel.
3   In 2011, IDC estimated the total amount of data worldwide at 1.8 zettabytes (1.8 trillion gigabytes). This total was growing at 40% annually. See *Digital Universe Study: Extracting Value from Chaos*, IDC, June 2011. For historical data growth patterns, see Tallon, P. P. and Scannell, R. "Information Lifecycle Management," *Communications of the ACM* (50:11), 2007, pp. 65-69.

purposes. In some instances, case law has prompted organizations to retain all electronic records indefinitely.[4] These three factors point to two potentially conflicting needs: the need to protect data against various technical and organizational risks and the need to enable greater use of data as a means of generating value.

This article describes how Intel addressed these conflicting needs through its information governance policy, which initially focused on protecting data but has evolved to a Protect-and-Enable approach that permits potentially risky but value-creating uses of data that were once discouraged. (The Appendix contains more information about the research conducted for this article.) As we argue in this article, the evolution of Protect-to-Enable at Intel reveals a level of maturity that allowed it to avoid a tendency—common across many IT units—to assert control by locking down access to data without a clear and compelling user justification. Protect-to-Enable balances the need to protect data and the need to make data available for exploratory or nontraditional uses. A key aspect of the approach is educating users and working with them to create structures and policies that set boundaries for what is allowed.

We have distilled Intel's experiences with information governance into five guidelines that can be applied by other organizations. A key lesson is to set information governance policies and structures that enable uses of data within a risk-aware environment.

Intel Corporation: An Overview

Although Intel has long been synonymous with microprocessor innovation, the company sees itself as more than a microprocessor company. Since the early/mid 2000s, Intel's strategy has evolved from a focus on designing and manufacturing microprocessors to delivering IT-based solutions that encompass aspects of microprocessor design but that can equally include software platforms and services. The bulk of Intel's revenues and profits continue to come from two of its five business units: PC Client Group (64% of sales; 89% of operating income) and Data Center Group (20% of sales; 35% of operating income).[5] Both groups encompass Intel's traditional microprocessor products that are used in desktop PCs, laptops and servers.

Like many organizations, Intel has struggled to manage its vast pool of data. By mid-2013, it was managing over 65 petabytes of data—150 times larger than the digital archives of the U.S. Library of Congress. This staggering volume of data has grown at 30% to 40% per annum for most of the last decade, easily swamping any percentage increases in sales and profitability. More importantly, the growth in data volume has occurred at a time when IT spending has declined from 3.5% of sales in 2004 to 2.4% in 2011. A particularly vexing aspect of this growth involves unstructured data,[6] which has grown faster than structured data and is more difficult to govern. Approximately 58% of Intel's data is unstructured engineering files that are used in product design. The complexity and scale of these files have increased with each new generation of microprocessor.

Intel's data challenges are further compounded by the global nature of its operations with seven fab plants in four countries (China, Ireland, Israel and the U.S.) and seven assembly, test and R&D facilities in five countries (Costa Rica, China, Malaysia, the U.S. and

---

4 Case law based on previous judgments comes from two main sources. The first involved a $1.57 billion judgment in 2005 against Morgan Stanley in favor of investor Ron Perelman arising out of his acquisition of Sunbeam. The court had requested specific data from the defendants only to be told, falsely, that no such data existed. The judge imposed punitive damages of $1.45 billion because of the defendant's failure to comply; the entire judgment was later reversed on appeal. In the second case, involving Zubulake v. UBS Warburg (2004)—a case alleging gender discrimination, failure to promote and retaliation—the judge criticized the defendant for insufficient e-discovery and destruction of documents. This case established the requirement for firms to produce data in court in a timely manner. Corporate legal counsel has reacted to these decisions by requesting that all data be retained indefinitely.

5 Sales and operating income data are based on Intel's 2012 10-K report, available via the U.S. Securities and Exchange Commission's Electronic Data Gathering, Analysis (EDGAR) archives (http://www.sec.gov).

6 Unstructured data refers to standalone files that exist separate from the applications that created them. Examples include digital media files (voice or video), CAD, office files (word processing or spreadsheets), email, web files and text documents. Because it is unstructured, such data is difficult to govern. For example, depending on the content, sender or recipient, some emails may need to be retained for significant periods of time. Metadata (data about data) is often used to describe unstructured data. In the case of email, this could be a subject line, although subject lines rarely describe with sufficient detail and accuracy the complete contents of each email.

Table 1: Information Governance Timeline

Year	Description of Significant Event
1992	Introduction of centralized IS organization (mainframe environment/WAN)
1995	PCs brought under central control
1998	Centralized applications consolidated in separate e-business group
2002	Compliance with Sarbanes-Oxley through information governance policies
2003	SQL Slammer virus impacts Intel
2004	E-business and infrastructure groups consolidated into one centralized IS organization
2005	Information protection and risk management efforts scaled up
2006	Manufacturing and factory systems brought under central control
2009	Governance inflection point reached—triggers the Protect-to-Enable approach
2010	Data analytics efforts launched

Vietnam).[7] Each of these seven countries operates a different set of directives and recommendations pertaining to privacy policies, data security, ownership, retention and use.

The Protect Era of Information Governance (2003-2009)

Like many organizations, Intel's initial approach to information governance was to implement policies and structures to lock down access to data. This approach arose from fears that critical systems and data, such as microprocessor designs and financial data, would be compromised if information governance policies were too lax.

The origins of information governance at Intel can be traced back to 1992 with efforts to consolidate and centralize network and mainframe IT, which had been distributed and locally managed by individual business units and sites. Table 1 provides a general timeline of the company's information governance initiatives from 1992 onwards. Desktop computing was centralized in 1995, although applications remained under local control until 1998 when a second IT unit (the e-business group) was created to bring applications under central control. This dual approach to IT management—one IT function managing all aspects of IT hardware with a separate IT function managing applications—remained in effect until 2004 when a single IT function with one CIO and a consolidated IT budget was formed to control all aspects of IT across all of Intel's business units and locations worldwide.

As part of the IT consolidation in 2004, Intel's Information Management (IIM) group initiated a Master Data Management (MDM) program. To support this effort, Intel was anxious to create a single set of MDM policies.[8] IIM needed a governance process that would engage business people across the enterprise in setting these policies. Intel also needed policies that would reflect different regional laws and the changing nature of Intel's activities. To that end, IIM created independent information governance boards for each of its six core master data areas: customer, supplier, location, item, worker and finance.

Two events led Intel to implement strong protectionist policies over data. First, the introduction of Sarbanes Oxley legislation in 2002 compelled the company to focus more on protecting financial transaction-level data. Second, and more importantly, in early 2003,

---

7   Intel's fab plants are considered chemical facilities under the 2011 Chemical Facility Anti-Terrorism Security Act. While the act articulates the need for physical access controls, it also specifies the need to protect key information.

8   Research shows that one of the first steps towards effective information governance is to use a consistent approach to data management across the enterprise. See Khatri, V. and Brown, C.V. "Designing Data Governance," *Communications of the ACM* (53:1), January 2010, pp. 148-152; and De Haes, S. and Van Grembergen, W. "An Exploratory Study into IT Governance Implementations and its Impact on Business/IT Alignment," *Information Systems Management* (26:2), March 2009, pp. 123-137.

Intel was impacted by the SQL Slammer virus, which infected Intel's internal networks through an employee's home-based remote network connection.[9] In response, Intel's CEO, Andy Grove, placed a corporate officer in charge of forming a cross-functional Safety and Security task force drawn from every major business unit and significant horizontal functions such as finance, HR, IT, environment, health and safety, corporate security and legal. This task force focused attention on the need to establish business continuity, risk and security efforts to guard against future attacks.

This marked the beginning of Intel's *Protect Era* of information governance. Efforts to lock down access to Intel's key IT assets were expanded in 2005 in advance of the consolidation and centralization of manufacturing and factory systems in 2006.

Intel was concerned with four types of risk:

- *e-discovery:* the ability to efficiently search, locate and recover key information within a given time interval, often as a result of a court order

- *Business continuity:* the ability to recover critical business operations with minimal long-term disruption

- *Compliance:* satisfying minimum standards for data retention, controls and access

- *Intellectual property:* cyber-attacks, theft or access violation.

One way Intel managed these risks was by regularly involving legal counsel in setting information governance policies. Legal counsel helped to interpret the ever-growing body of legal requirements and to set retention limits for various types of data, not only for e-discovery but in other cases where Intel IT might not fully grasp the legal ramifications of losing data.

Another manifestation of the Protect Era was Intel's approach to IT contractors. At the time, Intel adjusted its IT staffing needs through short-term contracting. To contain the risk that IT contractors might pose by introducing non-Intel devices onto Intel's internal network, it provided contractors with a fully loaded Intel laptop.

The resulting *Scorched Earth* approach to information governance prevented all unnecessary access to critical information assets. The result, however, was an increase in individuals devising policy workarounds. Management discovered that to complete certain tasks, engineers and other workers applied risky and often unauthorized workarounds that were within the letter, though not the spirit, of Intel's information governance policies. These workarounds increased the level of technical, organizational, reputational and financial risk, offsetting Intel's attempts to reduce risk. They also prompted a discussion among senior IT leaders on whether an all-consuming emphasis on controls and risk avoidance was likely to prove ineffective or to fail outright. As Intel's Chief Information Security and Privacy Officer noted about Intel IT's tendency to over-protect the company's data:

> *"When we just focused on protecting data, we had an over-control situation. We over-constrained and eventually generated more risk because people could figure out a way to go around the controls."*

Intel's desire to lock down and protect data resources also led to policies that often mandated retaining data on the most expensive storage devices, in some cases indefinitely. Users were oblivious to the cost of protecting data; chargeback models were not formally used to assign costs to specific users. As the volume of data grew, data management costs began to escalate. A member of Intel's Cloud Integration efforts observed how user behavior contributed to increased IT costs:

> *"Information growth and user behaviors are driving up IT infrastructure costs. The only way we can change that is to transition to a requirement-driven organization where if we understand the value of information and how to unlock that value, we can start to align value to data storage infrastructure."*

The Protect-to-Enable Era of Information Governance (2009 onwards)

Over time, the protectionist approach to information governance came to be seen

---

9 Intel was also affected by the Melissa virus (1999) and the Code Red virus (2001). Both viruses hit at a time when information governance was the responsibility of a 20-person corporate information security group.

as excessive, expensive, risk inducing and detrimental to Intel's long-term innovation efforts. By 2009, BYOD (bring your own device) was emerging as an accepted use of personal technology, and data analytics was gaining momentum in the practitioner literature. As a consequence, the initial approach to information governance was increasingly frustrating employees. Intel IT realized that its approach to information governance needed to evolve to become less restrictive and more accommodating of users' desires to use data or other information resources in new and nontraditional ways. This led Intel to adopt a *Protect-to-Enable* philosophy toward data management. Devised by Malcolm Harkins, Intel's Chief Information Security and Privacy Officer, Protect-to-Enable[10] was intended to generate business value through greater use of IT resources and data but within defined, quantifiable and tolerable risk limits. Harkins explained management's thinking on why governance policies from the Protect Era needed to be replaced:

> *"Our whole protectionist stance might have mitigated one form of risk but it increased other forms of risks. I think we would've had a lower cost and risk portfolio if we reacted differently. But, because I went Scorched Earth and reacted to what introduced harm to the company, I over controlled instead of appropriately controlling with a more nuanced and sophisticated approach. If we had allowed an unmanaged device on our network and had been more open to that dialog earlier, we would have designed controls earlier around being more progressive on BYOD. I made a risk decision. It was an imperfect decision but we let it stay in place too long and we never looked back. We would've been more innovative and we would've enabled the company more at a lower cost point if we had changed sooner. But, we didn't really make the switch to Protect-to-Enable until about 2009."*

Because innovation had long been a driving force behind Intel's success, management started to view the success of information governance in terms of whether it boosted innovation and reduced time to market. Cost savings were also a consideration. Therefore, an important element of Protect-to-Enable was communication with and education of users so they would learn how data management costs grow. An Enterprise Storage Architect at Intel observed:

> *"One of the biggest issues we had [before Protect-to-Enable] involved the lack of a proper cost model where we could charge users based on their use of storage. Users were able to dictate that they just wanted the fastest, the most smoking-hot storage irrespective of the criticality of the data. Under Protect-to-Enable, we now have a process where business needs and data classification [the value of data] are taken into account. A user is now given a cost associated with their storage usage and in a lot of cases you'll find that when they get the dollar figure, they realize that they don't need that level of performance or protection and can make do with a lower tier of storage. [However] if the user says they don't want to spend that amount of money [and] if Intel thinks the data is critical, data stewards will work with management to get that money. They're not going to allow users to keep critical data on a lower tier, just because users don't have enough money. The risk would be more than we want to tolerate."*

An important factor in Intel's transition to Protect-to-Enable was changing how users viewed data management. A member of Intel's Cloud Integration efforts highlighted an evolving role for data users:

> *"One issue we've seen as we move to an enablement model is the need to convince the various lines of business that they— not Intel IT—are the stewards of the data. They need to take ownership of their data including classifying its value and how it should be used before we can decide the mechanics of how the data will be managed. They know the value of their data better than we [Intel IT] do so we have to rely on them to provide that insight."*

An Enterprise Storage Architect outlined the governance process by which users or data stewards classify data according to its value and

---

10   For more information, see Harkins, M. *Managing Risk and Information Security: Protect to Enable*, ApressOpen, 2013.

how this outcome has implications for how data is managed:[11]

> "We require a review process that's primarily designed for business continuance. Users go through a questionnaire that finds the criticality of the data based on business needs and whether it is actually critical or just business-important. Based on that outcome, we can place data on an appropriate [storage] tier."

Responsibility for applying Protect-to-Enable across the enterprise was vested in Intel's Corporate Risk and Security Group (CRSG), whose responsibilities covered all electronic and paper records across all U.S. domestic and international locations, business units and employees. Its primary goals were:

1. Develop and administer guidelines for the retention and disposition of data; promote compliance by Intel employees and outside contractors with the guidelines and, where appropriate, with all international legal/regulatory mandates

2. Maintain corporate records in compliance with all legal obligations and preserve information if Intel is involved in litigation

3. Manage Intel information for which there is no legal or regulatory requirement or business need

4. Protect the privacy of information as required by law, regulation and Intel's privacy policies.

The CRSG assumed many of the strategic policy-making activities previously assigned to the governance boards that had overseen the MDM initiatives begun in 2004. The focus now was on evolving information governance to include data architects who understood the complexities of capturing and retaining massive amounts of data, and business leads (the ultimate owners of the data) who understood how data could be better applied for decision making.

While governance boards were still able to focus on technical MDM issues such as ensuring integration and consistent data standards across all critical applications, policy setting was vested in an Ethics and Compliance Oversight Committee within the Risk and Compliance function of Intel.

Given the strategic focus of this committee, representatives were drawn from critical support groups such as IT, HR, legal, business development and internal audit, with other representatives drawn from key business functions such as sales and marketing, manufacturing and product design. This committee continues to meet on a quarterly basis to review and, where necessary, propose information governance policies for each business unit and country. Business units are expected to conduct self-assessments periodically to determine if their policies are adequate for their needs and consistent with Protect-to-Enable, or if they should be expanded to account for new risks or changes within the business unit operating environment.[12]

The operational aspects of Intel's information governance are currently managed through technical IS roles and activities within Intel IT. Audits are used to check compliance with information governance rules and to propose high-level policy changes where necessary. Intel IT also conducts risk assessment and incident-response planning, activities that tie back to the oversight work of the Ethics and Compliance Committee. Operational activities also include data backups and disaster recovery planning because both activities are critical to maintaining an adequate level of user access to data for decision making.[13] Intel IT also plays a key role in monitoring information governance for compliance with domestic and international regulation. Internal audit and Intel's Security and Privacy Office monitor the internal and external

---

11 For further details on how Intel has structured its internal storage environment, see Bell, R. et al. "Solving Intel IT's Data Storage Growth Challenges," IT@Intel White Paper, January 2012; and Srinivasan, V., *The Evolution of Master Data Management at Intel: A Case Study of Finance Master Data*, April 2011 (http://www.dataversity.net/the-evolution-of-master-data-management-at-intel/).

12 For an analysis of how storage decisions reflect a balance of cost and risk, see Tallon, P. P. and Scannell, R., op. cit., 2007; Weber, K., Otto, B. and Österle, H. "One Size Does Not Fit All: A Contingency Approach to Data Governance," *Journal of Data and Information Quality* (1:1), 2009, pp. 1-27; Tallon P. P. "Understanding the Dynamics of Information Management Costs," *Communications of the ACM* (53:5), 2010, pp. 121-125; and Bell, R. et al., op. cit., 2012. For information on how to design data governance structures and policies that balance cost and risk, see Khatri, V. and Brown, C. V., op. cit., 2010.

13 The value of backups is assessed by two metrics: RPO (return point objective based, on the age of the last backup) and RTO (return time objective or the length of time needed to restore from backups).

landscapes for unfolding threats and other factors that could cause information risk to increase to unacceptable levels.

Intel considers external engagement to be a core element of information governance. Information about threats and vulnerabilities, best practices and benchmarking is widely shared with professional IS groups, such as the Information Risk Executive Council and the San Francisco Bay Area CSO Council.[14] Such engagements help to highlight external factors that could affect information governance at Intel.

How the Protect-to-Enable Approach Facilitates Analytics

Adopting a Protect-to-Enable approach to information governance in 2009 was fortuitous in light of the growing popularity of data analytics that emerged by the end of that decade. In 2010, Intel created a business intelligence (BI) data management group whose goal was to facilitate the collection, processing, retention and distribution of data needed for analytics. The desire among users for self-service BI and analytics capabilities forced Intel to reassess its information governance since data analytics often required access to data in other parts of the company. Part of the task of building analytics capabilities was allowing functional areas to see the value of their data when shared outside their immediate area. During the Protect Era, data stewards had restricted data access to within their functional areas. Analytics challenged that restriction by highlighting the potential for data access to add value in new ways.

Although MDM was no longer a limitation, users had no incentive to share data or even to publish its existence. Hence, some degree of data duplication existed before a firm-wide data analytics effort was established. The BI data management team began to function as a broker between users and owners of the data. Thus, the Protect-to-Enable approach moved information governance beyond the perspective of who owns the data to who can best use the data and what types of organizational value they might achieve from using data in new ways.

Measuring the Value of the Protect-to-Enable Approach

Intel is well known for its intense metrics-driven culture.[15] Because of the strategic role of data in Intel, the company monitors changes in costs (how data growth is impacting IT spending), in risk (how much value is lost if data is lost or compromised) and in value (how data is helping to improve firm performance).

The company evaluates the Protect-to-Enable approach in a similar way. Technical and business risk are consistently monitored through systems audits, while data storage costs are tracked to determine whether resources are being wasted or if spending targets are being met.[16] Value is measured in operational and strategic terms. A decline in data loss and security-related incidents that put data at risk indicates, at an operational level, whether data is being adequately protected. At a more strategic level, Intel uses agility as a measure of its ability to respond to market change and to design, test and deliver new products and services within an ever-shorter timeframe. For example, analytics has helped Intel to cut 25% off the time needed for chip design validation, thereby allowing it to launch products faster than its competitors and so maintain its lead in the microprocessor market. As an innovator, Intel examines whether data is enabling new growth opportunities and how data can benefit productivity and the overall effectiveness of its manufacturing operations.[17] In this way, Intel monitors both aspects of Protect-to-Enable—whether data is protected and whether data is enabling new opportunities for improving financial performance and market positioning.

Lessons Learned

Five key lessons emerged from Intel's experience. These lessons can be applied by other organizations as they seek to develop information governance—or to benchmark their current

---

14   See http://www.executiveboard.com/exbd/information-technology/it-risk/index.page.

15   Grove, A. S. *Only the Paranoid Survive: How to Exploit the Crisis Points That Challenge Every Company,* Crown Business Books, 1999.

16   Users are often under the mistaken belief that falling hardware costs allow for a greater level of data retention. In reality, however, for every dollar spent on storage hardware, three to seven dollars goes to non-hardware items such as labor and support expenses, software licensing and datacenter support costs.

17   Harkins, M., op. cit., 2013.

practices as they respond to the emergence of big data.

1. Eliminate Practices that Over-Govern Information

Management at Intel initially believed that data protection would minimize risk and drive success. There was little tolerance for data-driven risk, and it was relatively easy to implement policies to limit data access. When it became apparent that over-governance was instead driving up costs and increasing risk, management sought to identify and remove practices that were too restrictive. The Protect-to-Enable approach attempts to balance the need to protect data from various risk factors with the need to make data more accessible and available for decision making.

2. Educate Users about Data-Related Risk and Cost

Users may have minimal understanding of the costs associated with managing information. They tend to see information management costs purely in terms of purchase prices and fail to recognize that these costs are a fraction of the total cost of ownership. They also fail to appreciate the different types of risk and how their actions and behaviors can contribute to higher risk.

When organizations develop information governance policies and structures, users may not immediately understand why their actions are being regulated. They are focused on doing their jobs, not on the costs or risks of doing that job. Intel repeatedly educated its users as to why specific policies were in place. The goal throughout was to encourage users to work within the letter and spirit of Intel's information governance policy framework and then to trust them to act appropriately. Given the sharp rise in data costs, Intel also used a cost-allocation model to report data-specific costs to users (this is not a traditional chargeback model but rather an informational model).

Intel's approach to educating users about both the need for information governance and sensitivity to risks and costs can be summarized as:

- *Promoting personal responsibility:* since unstructured data accounts for a disproportionate share of storage costs and risk, users should be discouraged from unnecessary data hoarding and retention of files that are clearly of no future value, while encouraging more accurate assessment of storage needs.

- *Being proactive:* users will likely think that the data they use in their job is of the highest value to the organization and, therefore, appropriate for retention on tier 1 (the most expensive) storage. Data is rarely so valuable and, hence, users should be educated to quantify value and, where possible, to move their data to a lower storage tier.

- *Working with and not around policies:* information governance best practices are evolving, often in response to system failures or other adverse storage events. Since users may not always see the business logic behind each policy, information governance may be seen more as an obstacle than as an enabler of work. Bottom-up approaches to information governance, where each business unit or function is allowed to develop its own rules, fosters unnecessary complexity and inconsistency. If policy deviations are necessary, they should be decided by a corporate-wide function.

3. Collaborate with the Business to Design Information Governance

While responsibility for information governance may, by default, be assigned to the IT group, the scope of the governance policy requires the IS group to collaborate with representatives from key business units and functions. An important feature of the Protect-to-Enable approach at Intel was its collaborative nature; rules were not imposed on users in a way that might breed resentment. Instead, governance rules were co-created with business representatives with an eye to what is acceptable and appropriate for users in each business area.

4. Allow Exceptions to Global Policies to Meet Local Needs

Multinational corporations with significant operations outside the U.S. face additional challenges in managing information. A one-size-fits-all approach is unlikely to work since some policies may conflict with national regulations. For example, the E.U. considers IP addresses to be personal information, requiring additional layers of data protection or precluding the use and storage of this data.[18] Intel had to grapple with bring-your-own-device policies that limit device monitoring in some markets. It regularly assesses global policies and customizes local policies to meet unique regional or market needs.

5. Help Users Put a Financial Value on Data

Intel discovered that it is difficult to put a financial value on data, but users were still expected to work through a data-classification exercise to assign a value to their data. From this value, Intel IT computed the financial risk of losing the data and identified the steps needed to reduce risk to more tolerable levels. Over time, data stewards emerged as important stakeholders in helping to monitor changes in value. As the financial value of data shifted over time, Intel was able to migrate its data to appropriate storage tiers whose costs and service levels matched business requirements.

Concluding Comments

Forecasters predict that by the end of the next decade, the quantity of data under management in many organizations and business sectors will increase fifty-fold. Even the most conservative estimates of 40% annual growth imply a twenty-fold increase in the amount of data by 2020.[19] Like many organizations, Intel recognized the strategic value of its data and the need to carefully manage data in ways that allow value to be realized within tolerable risks limits. Rather than using information governance policies or structures to lock down data by controlling its use and access, Intel's use of Protect-to-Enable provided an evolving governance framework within which data was effectively shared and used within known and acceptable risk levels.

As data analytics becomes an increasingly important driver of sales growth and financial performance, managers will likely rely on information governance to determine what they can and cannot do with data. If organizations over-govern their data through bureaucratic and complex structures, or adopt policies that might be perceived by users as restrictive, costly and time-consuming, there is a risk that users will implement workarounds that completely bypass the governance structure. The risk of operating outside formal governance structures is that users may not readily grasp the risks of losing data. The point is not to avoid information governance entirely or to employ minimally intrusive policies but to educate users and work with them to create structures and policies that set boundaries for what is allowed.

Appendix: Research Methodology

In addition to information provided by Malcolm Harkins (the Intel co-author of this article), the academic co-authors conducted semi-structured interviews with seven subject matter experts at different levels in Intel IT. Interviews were tape recorded and transcribed. We then used content analysis on each transcript to reveal the history of information governance at Intel, the range of information governance practices in place, the risks these practices were intended to address and their overall level of success.

---

18   E.U. members have been given some flexibility in deciding how to apply the 1995 Data Protection Directive, which prescribes the need for information governance. The Article 29 Working Party (the E.U. Committee tasked with clarifying the directive) declared in 2007 and 2008 that IP addresses are personal data since they relate to an "identifiable person." Adding to the confusion, in 2008, a German court issued a contradictory ruling saying that IP addresses are not personal data and do not need to be protected as such. As of 2013, the E.U. is planning a comprehensive review of the 1995 directive to take account of new innovation and legal/compliance challenges. Outside the E.U., interpretations of the personal nature of IP addresses vary widely.

19   Reinsel, D. and Gantz, J. *The Digital Universe in 2020: Big Data, Bigger Digital Shadows, and Biggest Growth in the Far East*, IDC iView, December 2012 (http://www.emc.com/collateral/analyst-reports/idc-the-digital-universe-in-2020.pdf).

About the Authors

Paul P. Tallon

Paul Tallon (pptallon@loyola.edu) is Associate Professor, Frank J. DeFrancis Chair in Information Systems and Director of the Lattanze Center for Information Value at the Sellinger School of Business, Loyola University, Maryland. Previously, he was an IT auditor and accountant with PricewaterhouseCoopers. He has published over two dozen papers in journals including *MIS Quarterly*, *Journal of Management Information Systems*, *European Journal of Information Systems*, *Journal of Strategic Information Systems* and *Communications of the ACM*. He won the 2007 *Journal of Management Information Systems* paper of the year. His research interests cover the economic impacts of IT, strategic IT alignment, real options, IT risk and the economics of information management.

James E. Short

James Short (jshort@ucsd.edu) is Lead Scientist, Center for Large-Scale Data Systems, at the San Diego Supercomputer Center, University of California, San Diego. Previously, he was on the faculties of the MIT Sloan School of Management and the London Business School. His work has been published in *The Accounting Review*, *MIS Quarterly*, *Journal of Management Information Systems*, *Journal of Management Studies*, *International Journal of Communications* and *Sloan Management Review*. His research interests cover the value of data and information growth, information transparency and financial performance, and data metrology and measurement.

Malcolm W. Harkins

Malcolm Harkins (malcolm.harkins@intel.com) is Vice President and Chief Information Security and Privacy Officer at Intel Corporation. In this role, he is responsible for managing the risk, controls, privacy, security and other related compliance activities for all of Intel's information assets, products and services. He was previously Chief Information Security Officer and has held several other roles at Intel spanning finance and procurement. He was recognized by *Computerworld* magazine as one of the top 100 IT Leaders for 2012. He is the author of several Intel white papers and, most recently in 2013, his first book, *Managing Risk and Information Security, Protect to Enable*, was published by ApressOpen.

Crowdsourcing: How to Benefit from (Too) Many Great Ideas

This article focuses on how companies can cope with the enormous volume and variety of data (big data) that is acquired on crowdsourcing platforms from the worldwide community of Internet users. We identify the challenges of implementing crowdsourcing platforms and show how CIOs and other organizational leaders can build the absorptive capacity necessary to extract business value from crowdsourced data.[1,2]

Ivo Blohm
University of St. Gallen
(Switzerland)

Jan Marco Leimeister
Kassel University (Germany),
University of St. Gallen
(Switzerland)

Helmut Krcmar
Technische Universität
München (Germany)

The Power and Challenges of Crowdsourcing

Digitization and the Internet have empowered firms to tap into the creative potential, knowledge and broad-based experience of a huge crowd of contributors. For instance, the gold producer GoldCorp made its geographical databases available to the public and offered a prize for anyone who could tell it where to find gold. The results of this open call enabled GoldCorp to increase its gold production from 53,000 to 504,000 ounces a year while it cut production cost from $360 to $59 per ounce. As a consequence, the value of GoldCorp increased from $100 million to $9 billion.[3] InnoCentive provides an online platform that enables organizations to present engineering problems that they are unable to solve in-house to a community of hobby scientists. On average, InnoCentive's hobby scientists solve 30% of these problems.[4] Similarly, TopCoder, a pioneer of community-driven open innovation, provides a community of software coders who frequently produce more effective software algorithms at lower cost than traditional software creation approaches.[5] All three are examples of *crowdsourcing* and

1 Cynthia Beath, Jeanne Ross and Barbara Wixom are the accepting senior editors for this article.

2 An earlier version of this article was presented at the pre-ICIS SIM/MISQE workshop in Orlando, Florida, in December 2012. We thank Sabine Matook and John Mooney, who provided invaluable feedback on a previous version of this article, and two anonymous reviewers for their very helpful comments.

3 For more details on GoldCorp's crowdsourcing approach, see Tapscott, D. and Williams, A. D. *Wikinomics: How Mass Collaboration Changes Everything*, Portfolio, 2008.

4 For more information on InnoCentive, see Jeppesen, L. B. and Lakhani, K. R. "Marginality and Problem-Solving Effectiveness in Broadcast Search," *Organization Science* (21:5), 2010, pp. 1016-1033.

5 For more on TopCoder, see Lakhani, K. R. et al. "Prize-Based Contests Can Provide Solutions to Computational Biology Problems," *Nature Biotechnology* (31:7), 2013, pp. 108-111.

illustrate how companies have successfully managed to absorb the volume and variety of crowdsourced data to create business value.

The fundamental idea of crowdsourcing is that a crowdsourcer (which could be a company, an institution or a non-profit organization) proposes to an undefined group of contributors (individuals, formal or informal teams, other companies) the voluntary undertaking of a task presented in an open call. The ensuing interaction process unfolds over IT-based crowdsourcing platforms.[6] The power of crowdsourcing lies in aggregating knowledge from a multitude of diverse and independent contributors. Crowdsourcing enables crowdsourcers to obtain solutions that are beyond the boundaries of their established mindset.[7]

There are two types of crowdsourcing: *tournament* and *collaboration*. In collaboration-based crowdsourcing, contributors create collectively a common solution (e.g., an entry in Wikipedia). Such solutions are the result of many small contributions that individually have minimal value. By contrast, tournament-based crowdsourcing involves the submission and collection of independent solutions (e.g., ideas, prototypes, business plans). The crowdsourcer selects one or a few of the contributions in exchange for financial or non-financial compensation. Tournament-based and collaboration-based crowdsourcing can be combined—for example, by using collaborative evaluation and improvement of individual contributions in tournament-based crowdsourcing.[8]

Successful crowdsourcing platforms easily attract tens of thousands of contributors who create a huge volume of data of high variety. Crowdsourcers are often overwhelmed by this big data and find that creating business value from it is a time-consuming, resource-intensive and costly challenge, in particular if they lack the

capabilities and routines for making sense of, and then using, crowdsourced data. For instance, IBM employed 50 senior executives for several weeks to evaluate all 50,000 ideas that were submitted by its employees for further developing IBM products in one of its "Innovation Jams."[9] Similarly, it took Google almost three years and 3,000 employees to condense and translate the 150,000 proposals submitted to its "Project 10 to the 100" to 16 idea clusters, to evaluate these idea clusters, to develop appropriate projects for the most promising idea clusters and, finally, to start the projects.[10]

This article addresses how companies can cope with the enormous volume and variety of big data acquired via Internet-based crowdsourcing platforms. Based on our analysis of three crowdsourcer firms (one of which provides two crowdsourcing platforms), we show how CIOs and other organizational leaders can develop an effective *absorptive capacity* to enable them to generate knowledge and value from crowdsourced data.

Table 1 summarizes the characteristics of the crowdsourcer cases we investigated and their crowdsourcing platforms. Names of the involved companies and their platforms are disguised to maintain confidentiality. (Details of the research conducted for this study are in the Appendix.) Based on our case research, we present six recommendations for organizations seeking to improve their crowdsourcing effectiveness.

Implementing Crowdsourcing: The Case of BetaCorp's IdeaZone

BetaCorp's IdeaZone highlights the typical challenges of establishing crowdsourcing platforms. This case illustrates that the seeds of most of the challenges of dealing with crowdsourced data are planted in the early stages of a platform.

IdeaZone was started as a pilot project in 2009 and was later institutionalized as one of BetaCorp's standard programs for customer interaction. BetaCorp is a multinational software

---

6   For a sophisticated definition of crowdsourcing, see Estellés-Arolas, E. and González-Ladrón-de-Guevara, F. "Towards an Integrated Crowdsourcing Definition," *Journal of Information Science* (38:2), 2012, pp. 189-200.

7   The benefits of crowdsourcing as a problem-solving approach are discussed by Afuah, A. and Tucci, C. "Crowdsourcing as a Solution to Distant Search," *Academy of Management Review* (37:3), 2012, pp. 355-375.

8   The different modes of crowdsourcing are discussed in Zhao, Y. and Zhu, Q. "Evaluation on crowdsourcing research: Current status and future direction," *Information Systems Frontiers*, 2012.

9   Bjelland, O. M. and Wood, R. C. "An inside View of IBM's 'Innovation Jam,'" *MIT Sloan Management Review* (50:1), 2008, pp. 32-40.

10   http://googleblog.blogspot.de/2009/09/announcing-project-10100-idea-themes.html.

Table 1: Crowdsourcers and their Crowdsourcing Platforms

Company	AlphaCorp	BetaCorp		GammaCorp
Crowdsourcing Platform	Brainstorm	IdeaZone	Steampunk	Planet CoCreate
Number of Employees	500	61,000		7,000
Target Group of Crowdsourcing Platform	Employees and customers	Customers	Employees	Customers
Focus of Crowdsourcing Platform	Developing innovative products	Developing and testing ideas for creating new products	Developing prototypes and products	Collecting concepts for improving products
Dominating Type of Crowdsourcing	Collaboration	Tournament	Collaboration	Tournament
Contributors (platform users)	35,000	10,000	200	2,000

manufacturer with a mature and established product-oriented organizational structure. Ten years ago, BetaCorp began to build a global ecosystem of online communities. Today, this ecosystem has more than three million users and consists of a plethora of forums, wikis, blogs and social networks.

BetaCorp launched IdeaZone for two reasons. First, employees requested a centralized channel for collecting customer and end-user feedback from the community ecosystem. Prior to IdeaZone's launch, employees had to extract customer feedback from various sources and could not communicate with customers directly without organizing user group meetings. Thus, IdeaZone was started as an open channel to collect and discuss new ideas for improving BetaCorp with customers and end-users. Second, IdeaZone was designed to have various evaluation functionalities not available on existing platforms. The aim of IdeaZone was to help BetaCorp employees test and prioritize customer and user feedback.

IdeaZone was started as a grassroots project by the managers of the community ecosystem. The entire project was initially independent of BetaCorp's IT department and started without official approval; corporate concerns about intellectual property, IT security and public

relations (i.e., public visibility of negative feedback) probably would have prevented the launch of IdeaZone. Top management was not involved until the IdeaZone platform was rolled out. The platform was licensed from a third-party provider IdeaZone was integrated into BetaCorp's community ecosystem IT infrastructure.

After launching IdeaZone, the major challenge for its initiators was to create internal awareness of the platform and to overcome the reluctance of employees to participate in it. Due to the grassroots character of the project, some employees feared that any time and effort they invested in using the platform might not be recognized or rewarded. The platform initiators overcame this challenge by actively recruiting a group of employees who were highly motivated to use crowdsourcing (most of them were so-called "digital natives") and supporting them in the execution of successful flagship projects stemming from crowdsourced ideas.

Based on these early successes, IdeaZone quickly gained momentum. The standalone nature of the platform allowed BetaCorp to adapt it to the needs of its external contributors and BetaCorp employees. During this phase, IdeaZone was developed into a platform on which employees from most business units posted tasks for the crowd of external contributors in a

Figure 1: Absorption Challenges

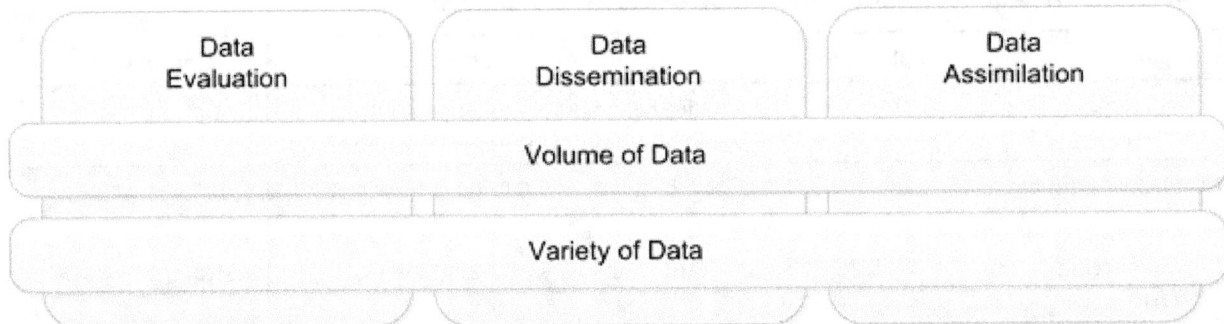

challenge-like fashion. Once the adoption of the platform and the various processes by which the different business units used it stabilized, the platform was technically integrated into the community ecosystem and BetaCorp's IT infrastructure. Today, IdeaZone is run by BetaCorp's IT department as a shared service. As such, IdeaZone serves as an open channel to all business units.

When the initiators launched IdeaZone, they envisioned implementing 10% of the crowdsourced contributions. So far, about 12,000 contributions have been collected (several times more than expected), of which 5% have been delivered or are under implementation for 32 different BetaCorp products. Business units systematically analyze and aggregate the contributions to identify new trends and shifting customer requirements, thus improving the strategic flexibility of BetaCorp. Employees also submit their own early-stage ideas to IdeaZone to see how contributors react and how their ideas stack up against other contributions. Business units now consider IdeaZone as a primary means of communicating with customers.

Challenges of Absorbing Crowdsourced Data

Having built up a successful crowdsourcing platform such as BetaCorp's IdeaZone, effectively exploiting crowdsourced data remains a challenge. In particular, the volume and variety of crowdsourced data inhibit the ability of companies to evaluate, disseminate and assimilate it, as depicted in Figure 1.

Volume of Data

Crowdsourcing platforms can collect data from very large numbers of contributors. The collected data is of three types:

1. *Contributions:* ideas, prototypes or business plans that are suggested solutions for the posted task

2. *Metadata:* examples include evaluations, comments or tags for individual contributions

3. *Data on contributors:* for example, personal characteristics, activity, preferences, evolving social networks on the platform and the quality of contributions (based on peer feedback).

The challenge of crowdsourcing is not only the sheer volume of data generated, but also the rate at which that data is created. For instance, AlphaCorp's Brainstorm gathered more than 8,000 contributions during the first weekend after its roll-out.

Variety of Data

To maximize contributor creativity, crowdsourcers usually do not put any format and structure constraints on potential contributions. This is particularly the case for tournament-based crowdsourcing where crowdsourcers are looking for existing solutions and well-developed prototypes that will address a given problem. As a consequence, contributions may lack focus and specificity. Contributions for the same task may differ dramatically in format, ranging from text-based descriptions to graphic visualizations to fully developed prototypes. Further, open calls attract contributors with highly diverse backgrounds who may propose very different solutions for the same task. Thus, the quality of contributions may vary considerably; typically there are a very few "extreme solutions" of high value and many solutions of moderate or low value.[11] Low-quality contributions tend to be highly ambiguous and unspecific, containing very little information the crowdsourcer can act on.

Data Evaluation

For crowdsourcers, the volume and variety of contributed data complicate idea evaluation. While the high volume makes it impossible to evaluate all the data manually, its variety inhibits automation of the evaluation task. Moreover, crowdsourcers may lack sufficient background knowledge to evaluate the data in all its richness. As a consequence, crowdsourcers apply various evaluation mechanisms, such as asking contributors to rate the quality of contributions from others. However, the design and use of such evaluation mechanisms is highly challenging. For instance, poorly designed rating scales can produce close to random results.[12]

Additionally, crowdsourcers are sometimes not aware that evaluation of contributions is also a very time-consuming task for contributors. As a consequence, many contributors evaluate only a small number of contributions, which means that many contributions do not have enough evaluations to be reliable. This problem is compounded if evaluation scales are misused by contributors. For instance, AlphaCorp recognized that contributors assessed contributions positively even though they apparently had no opinion about them. Faced with just a binary scale (i.e., thumbs up/thumbs down), some contributors positively assessed all 15,000 contributions on the platform. Thus, evaluation mechanisms may produce highly ambiguous results that cannot be interpreted clearly.

Data Dissemination

Disseminating crowdsourced data involves identifying and selecting appropriate employees and business units that will be responsible for assimilating the data and subsequently implementing the idea. This is an important step in the absorption of crowdsourced data as inappropriate recipients may not understand the data or may just ignore it. Due to its variety, data from crowdsourcing platforms might be of relevance to several business units. Thus, finding the right employees for each promising contribution is challenging. Some individuals may not be open to crowdsourced ideas (the "not invented here problem"). Sometimes, employees might not feel responsible for using the crowdsourced data. And some employees might become overwhelmed and suffer from information overload.

Data Assimilation

The assimilation of crowdsourced data is the process of transforming crowdsourced data into valuable information the crowdsourcer firm can act on by combining the data with the existing knowledge of the firm. The process involves the firm in developing concepts or business cases for commercializing crowdsourced ideas. At GammaCorp, for example, each contribution that is selected for implementation goes through the standard resource-allocation process involving analysis of technical and economic feasibility, strategic fit and an estimate of potential revenues. However, given the characteristics of crowdsourced data, the assimilation process may be arduous and lengthy. Selected contributions and their related data may have to be aggregated, translated and modified so they can be assessed against internal prerequisites such as corporate strategy and resource constraints.

Developing Absorption Capabilities for Crowdsourced Data

To deal with crowdsourced data and the associated absorption challenges, crowdsourcers need to build *absorptive capacity*—the capability to transform crowdsourced data into knowledge and business value. Thus, absorptive capacity depends on a company's processes for evaluating, disseminating and assimilating crowdsourced data so that it can create business value. To build absorptive capacity for crowdsourcing and overcome the absorption challenges, companies

---

11 For a deeper discussion of extreme solutions, see Jeppesen, L. B. and Lakhani, K. R., op. cit., 2010.
12 For a more detailed discussion, see Riedl, C., Blohm, I., Leimeister, J. M. and Krcmar, H. "The Effect of Rating Scales on Decision Quality and User Attitudes in Online Innovation Communities," *International Journal of Electronic Commerce* (17:3), pp. 7-37, 2013.

Table 2: Crowdsourcing Absorption Capabilities

Capability	Description
Platform Design	Designing a crowdsourcing platform that maximizes the quality of the contributions
Filter Design	Creating filtering processes that enable crowdsourcers to eliminate weak contributions early
Organizational Integration	Integrating crowdsourcing platforms into the organizational processes and structures of the crowdsourcer
Information Exchange	Managing the information exchange between contributors and the crowdsourcer's employees
Community Building	Attracting a critical mass of contributors and integrating them into a community of contributors

need to develop five distinct capabilities, which are summarized in Table 2.[13]

Platform Design

Crowdsourcing platforms shape how contributors generate contributions and how they interact with other contributors and contributions. Platforms structure the creative processes of the contributors and define the structure, format and quality of crowdsourced data. Thus, a well-designed platform appropriately supports the crowdsourcer, improves the value of crowdsourcing and eases the dissemination and assimilation of crowdsourced data by mitigating the challenges of its volume and variety.

AlphaCorp, for example, asked its contributors to provide new ideas for improving its products and found it could greatly improve the quality of contributions by asking contributors to provide both need and solution information. Need information describes wishes and requirements. Solution information describes how a need could be fulfilled or a problem solved.[14] AlphaCorp includes "rationales" and "solutions" on its crowdsourcing platform. Rationales contain a problem description, whereas solutions cover possible implementations that will solve the problem (see Figure 2). By entering need and solution information separately, contributors think not only about the problems but also about how to fix them. This separation helps contributors to present their contributions in a manner that better suits the mindset of the crowdsourcer's employees, who are predominantly interested in how existing problems and customer needs can be solved most effectively. Thus, the crowdsourced idea can be better understood, leading to improved evaluation, dissemination and assimilation.

Promoting collaboration among contributors is also important for improving the quality of contributions. Data pools, in which all contributions and their related metadata are visible to all contributors, enable contributors to explore, comment on and edit existing contributions (e.g., via wikis). AlphaCorp and BetaCorp supported collaboration by connecting potential collaborators during the contribution process. AlphaCorp's Brainstorm platform pools contributions so that contributors can add their rationales to existing solutions (see the Filter Design subsection below for more details). This pooling sparks intense discussions about the merits of the different solutions. Similarly, BetaCorp uses web conferences to enhance collaboration between contributors. For example, BetaCorp's Steampunk platform allows contributors to host brainstorm sessions in which peers collaboratively improve contributions. Both approaches build discussion groups and teams around single contributions, thus inducing collaboration-based crowdsourcing. Collaboration-based crowdsourcing improves the quality and understandability of crowdsourced data which, in turn, reduces the evaluation, dissemination and assimilation challenges.

---

13 An excellent discussion of how such capabilities form absorptive capacity can be found in Jansen, J. J. P., Van den Bosch, F. A. J. and Volberda, H. W. "Managing Potential and Realized Absorptive Capacity: How Do Organizational Antecedents Matter," *Academy of Management Journal* (48:6), 2005, pp. 999-1015.

14 A more detailed discussion of need and solution information can be found in Von Hippel, E. *Democratizing Innovation*, MIT Press, 2005.

Figure 2: AlphaCorp Brainstorm Toolkit

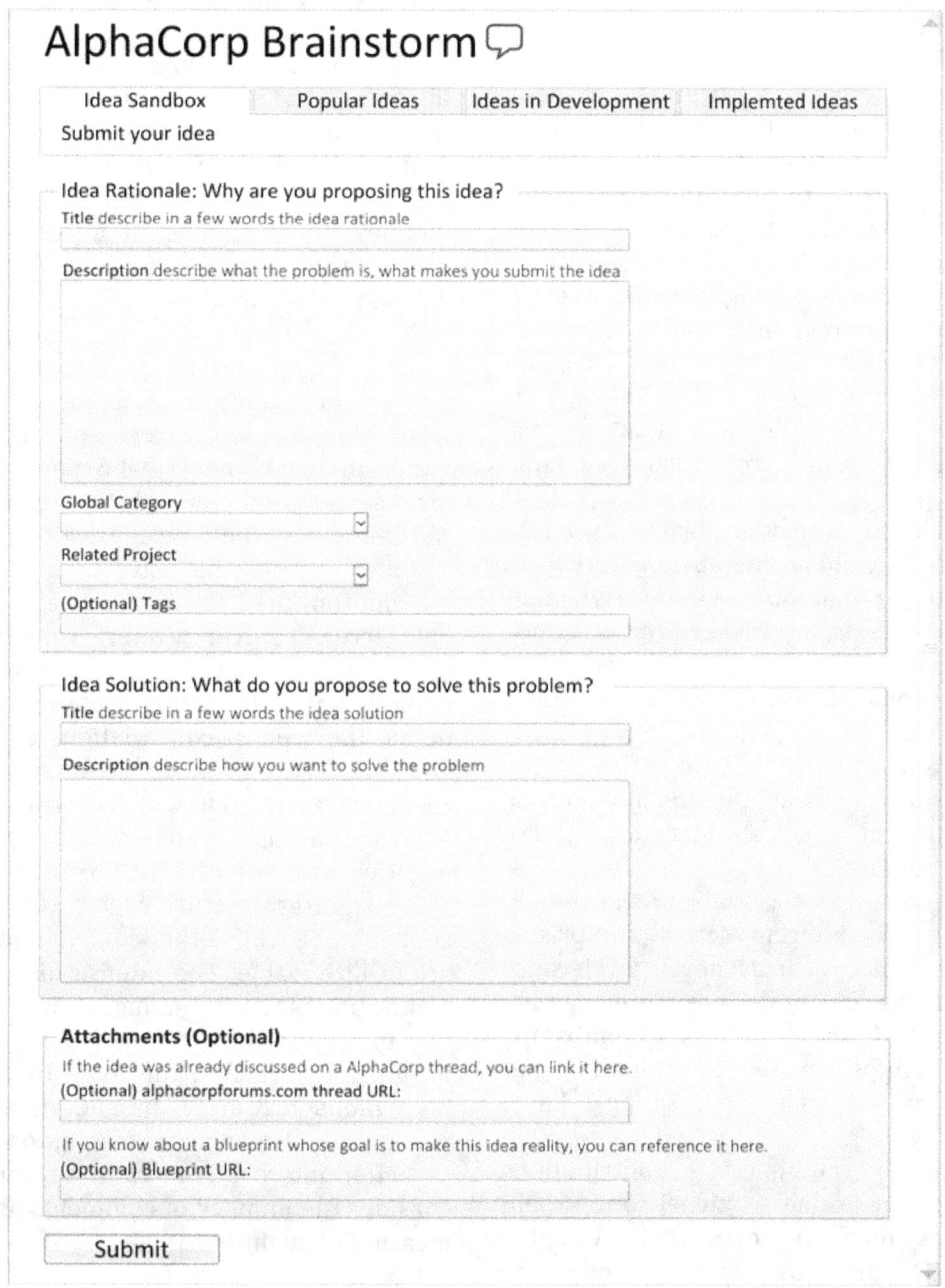

Finally, crowdsourcing platforms must motivate contributors. AlphaCorp's platform not only spurs collaboration, but also a sense of competition among contributors striving for the best solution. As a consequence, contributors often submit additional material that improves their contributions and increases their reputation and value. Other effective mechanisms for motivating contributors include gamification features such as rankings and point systems.[15] BetaCorp awards points for each contributor activity; point rankings show the most active contributors. BetaCorp uses this approach to engage contributors in making evaluations and

---

15 Gamification is the enrichment of products, services and information systems with game-design elements to positively influence motivation, productivity and behaviors of users. For more details, see Blohm, I. and Leimeister, J. M. "Gamification. Design of IT-Based Enhancing Services for Motivational Support and Behavioral Change," *Business Information Systems Engineering* (5:4), 2013, pp. 275-278.

comments so that existing contributions can later be evaluated more reliably.

Filter Design

To improve their absorptive capacity, crowdsourcers need to establish filter mechanisms that help them evaluate contributions early during the absorption process. Filters enable them to focus their limited resources on the most promising contributions and thus leverage the effectiveness of crowdsourcing. Good filter design identifies reliable contributions and aggregates them for evaluation. Filter design involves developing technical and organizational evaluation and aggregation mechanisms for crowdsourced data.

However, poorly designed filter mechanisms are vulnerable to evaluation biases (such as a contributor giving a thumbs-up rating to all contributions) that may lead to erroneous selection decisions. Crowdsourcers must therefore carefully design evaluation mechanisms. For instance, AlphaCorp improved the quality of the evaluation process by introducing an additional neutral rating option instead of just a thumbs-up/thumbs-down rating so contributors could indicate that they had not read a contribution.

Research shows that crowdsourcers should use multi-criteria filtering scales comprising several dimensions, such as novelty, relevance and feasibility, as such scales are more accurate than single-criteria scales such as a thumbs up/thumbs down. Approximately 20 evaluations per contribution are necessary for creating reliable quality rankings.[16]

Duplicates (multiple contributions that contain the same content) and spam (contributions that do not contain valuable information for the crowdsourcer) hinder the effective evaluation of contributions. Both detract attention from the most promising contributions and diminish the effectiveness of filtering mechanisms.[17] AlphaCorp eliminates duplicates and spam by using a multi-step procedure for new submissions. Each new

contribution is initially subject to a duplication check. Contributors have to enter the titles of their contributions, which are automatically compared with existing contribution rationales. If there is a match, contributors are invited to add their solutions to those. Next, new contributions enter a "sandbox." To get out of the sandbox, two other contributors must confirm the novelty and value of a contribution (i.e., confirm that it's not a duplicate or spam). Then, validated contributions are evaluated with a rating scale (using a simple on-screen slider). Finally, inappropriate contributions can be marked for further investigation by moderators, who may be employees of AlphaCorp or important contributors who participate in the management of the crowdsourcing platform (see the Community Building subsection below).

Filter mechanisms have to be accompanied by appropriate selection rules. Such rules help crowdsourcers filter contributions by systematically selecting the most promising among many thousands. Selection rules thus define the type of contributions crowdsourcers will see. For instance, BetaCorp's selection rules extract "polarized" contributions—those receiving a high share of both positive and negative evaluations. Such contributions have some features that are highly valued by some contributors, while other contributors think they are of little value. The intense discussions such contributions spur among contributors mean that they will likely provide some really valuable insights. To evaluate this type of contribution, both AlphaCorp and GammaCorp not only use ratings, but also analyze contributors' comments to better interpret the ratings. Moreover, they consider the number of comments as an implicit measure of quality.

Organizational Integration

It is important that crowdsourcers integrate their crowdsourcing platforms into their organizational processes and structures. Dissemination and assimilation require that crowdsourced data be explicitly integrated into the working processes of employees. To achieve this, responsibilities for the crowdsourcing platform and its data must be clarified. For instance, BetaCorp employees who want to post tasks on IdeaZone have to commit to devoting at least two working hours per week to responding

---

16 For a more detailed discussion, see Riedl, C., Blohm, I., Leimeister, J. M. and Krcmar, H., op. cit., 2013.

17 An extraordinary discussion of challenges of idea evaluation is provided by Di Gangi, P. M., Wasko, M. M. and Hooker, R. E. "Getting Customers' Ideas to Work for You: Learning from Dell How to Succeed with Online User Innovation Communities," *MIS Quarterly Executive* (9:4), 2010, pp. 213-228.

to crowdsourced contributions, commenting on them or assimilating them by readjusting internal projects based on the new insights. This commitment emphasizes that working with the platform is the only way of benefitting from it, and that it is employees' responsibility to actively make use of crowdsourced data. Similarly, GammaCorp changed the job profiles of some marketing and R&D employees to include responsibilities for monitoring the crowdsourcing platform and implementing platform-derived ideas.

Integrating a crowdsourcing platform into organizational processes and structures means that most employees consider the platform as just another of the information systems they use in their daily jobs. To minimize the effort of using a crowdsourcing platform, it could be integrated into existing information systems. For instance, GammaCorp integrated its crowdsourcing platform into its central knowledge management system. Thus, contributions and insights stemming from crowdsourced data become part of the corporate "collective memory." This improves dissemination and assimilation of crowdsourced data and blurs the distinction between internal and external contributions, increasing the acceptance of crowdsourced data.

Assimilation is also supported by training sessions on the value of crowdsourcing. During the launch of IdeaZone, BetaCorp invested substantial time in convincing critical employees of the benefit of expressing their ideas on the platform. Employees had be taught how to deal with a significantly expanded set of social interactions. Many employees may be unsure at first about how to handle negative feedback from other contributors, but they can be coached to be open-minded and understanding.

To encourage employee participation, efforts must be made to mitigate any perceived career risks from engagement in crowdsourcing. This could involve using C-level executives as promoters or allowing time to work on projects triggered by crowdsourced data. At BetaCorp, employees working with Steampunk receive "Thank God it's Friday" time that can be freely allocated to their own projects. Similarly, GammaCorp provides a fixed amount of resources (up to one person year for each posted task) for turning crowdsourced ideas into action.

Information Exchange

Managing the exchange of information between the crowdsourcer's employees and the contributors is crucial for the effective dissemination and assimilation of contributions. Providing contributors with feedback on their contributions is key to long-term success and to the development of future contributors. Thus, AlphaCorp and BetaCorp actively manage contributor expectations by, first, creating realistic expectations on the implementation of contributions. Additionally, they use "status signaling mechanisms" to track the stage of a contribution in the absorption process (e.g., "under review" or "implemented"). AlphaCorp also posts "developer comments" provided by the specific developer working on the implementation of a contribution. These comments are highly visible on the platform; they are viewed by many contributors and highlight AlphaCorp's engagement. Additionally, both these crowdsourcers make active use of blogs to explain certain decisions (e.g., why a highly popular contribution has not been implemented) and to be transparent.

Similarly, the crowdsourcer's employees must be actively encouraged to engage with the platform and crowdsourced data. AlphaCorp's and GammaCorp's platform managers aggregate information from contributors and present it in internal workshops. They use various internal communication channels to promote crowdsourced ideas and related contributions. For instance, they use newsletters or "crowdsourcing reports" in which the most promising contributions are highlighted. BetaCorp has "crowdsourcing mentors" who organize local events to promote Steampunk and the most promising contributions emerging from the platform. These mechanisms support assimilation of crowdsourced data by ensuring that it is recognized as valuable, as well as improving its evaluation and dissemination.

Finally, AlphaCorp and GammaCorp created opportunities for direct exchange between their employees and contributors by inviting contributors to company events. They also used various IT-related communication channels to extend the events' reach, including web conferences and social media such as

Twitter.[18] Similarly, BetaCorp actively uses web conferences for integrating contributors during the assimilation process. These conferences help resolve complex problems that may not have been solved by comment-based discussions because of the variety of crowdsourced data. As a by-product, contributor motivation is increased, and because there are more opportunities for discussing and refining crowdsourced contributions, the process of assimilating data is enhanced.

Community Building

Crowdsourcing relies on a community of contributors with self-organizing social structures. This requires the continuous acquisition of new contributors and the creation of common values and norms, especially empathy and trust. A strong community improves absorptive capacity because fewer resources will be needed for platform management, as these activities are performed in a self-organizing manner by the contributors themselves.

The key to building a vibrant crowdsourcing platform is to attract a critical mass of contributors so that a self-energizing cycle develops. The more contributors participating, the more attractive a crowdsourcing platform becomes for others, possibly leading to more crowdsourced contributions and a healthy inflow of the evaluations, comments or tags that are necessary for data evaluation. Attaching platforms to existing online communities is a highly successful practice, as demonstrated by BetaCorp's IdeaZone. BetaCorp integrated "feedback buttons" into some of its products, which also led to the acquisition of new contributors.

To tie new contributors to the crowdsourcing platform and to stimulate ongoing participation, contributors have to be integrated emotionally into the platform. To achieve this, BetaCorp's Steampunk invested significant resources in building a vital community of contributors, including the creation of a shared culture with which contributors can identify. With its rebellious attitude, Steampunk targets BetaCorp

Figure 3: BetaCorp Steampunk Vision

We Steampunks think that innovation @ [BetaCorp] is real and working. What's not working is getting resources and bringing innovation out on the market. Those steps are totally broken. We now take this into our own hands. Together we want to improve the way innovation is done, that projects can be proposed by each Steampunk, that you can find or be a resource (beside your regular projects), and that innovative stuff finally gets on the market. We give you the opportunity to propose and participate in innovative projects that allow you to make a deep dive, take your time, that you own and have an impact with, that allow you to build credibility in other areas than your daily job allows, and that aren't taken away by yet another reorg, change of manager, project cancellation or other reason. Punk!

employees who want to "fight against the encrusted structures" in a mature, multinational organization. This "outlaw" attitude resonates throughout the design of the platform and is encapsulated in a character who features prominently on the platform's homepage (see Figure 3). As a result, Steampunk contributors quickly develop and internalize a shared culture, one that helps develop a better understanding of the proposed tasks.[19] This shared culture helps BetaCorp actively set an agenda for the type of contributions it is seeking.

Emotional integration is also enhanced by building up self-organizing structures that enable contributors to engage actively in the management of the community. This integration fosters social ties among contributors as well as between contributors and employees of the crowdsourcer, which facilitates information exchange, and data evaluation and dissemination. To support these processes, AlphaCorp and GammaCorp define specific roles and rights for contributors (see Table 3). This hierarchy of roles clarifies what activities each type of contributor is allowed to perform. The more active and more recognized contributors become, the higher they rise in the hierarchy.

---

18   For an extensive description of how social media can improve socialization processes, see Jarvenpaa, S. L. and Tuunainen, V. P. "How Finnair Socialized Customers for Service Co-Creation with Social Media," *MIS Quarterly Executive* (12:3), 2013, pp. 125-136.

19   An excellent description of how a shared culture among knowledge management platforms can be developed is given in Teo, T. S. H., Nishant, R., Goh, M. and Agerwal, S. "Leveraging Collaborative Technologies to Build a Knowledge Sharing Culture at HP Analytics," *MIS Quarterly Executive* (10:1), 2011, pp. 1-18.

Table 3: Roles and Rights Hierarchy of AlphaCorp's Brainstorm

Role	Rights
User	Submit, evaluate and comment on ideas
Idea Reviewer	Validate new ideas in (+ User rights)
Moderator	Move ideas, change title or content of ideas (+ Idea Reviewer rights)
Developer	Provide developer comments (+ Moderator rights)
Administrator	Delete and ban users and distribute roles (+ Developer rights)

The possibilities for increasing their rights are highly motivating for many contributors and thus facilitate the emotional integration of contributors.

Recommendations for Building Effective Absorption Capacity for Crowdsourcing

Based on the absorption challenges and the capabilities that overcome them identified in our case studies, we provide six recommendations for CIOs and other organizational leaders as they set about building absorptive capacity for crowdsourcing.

1. Adopt a Broad Definition of Crowdsourcing Success

CIOs and organizational leaders should expect the business value of crowdsourcing to be multidimensional. Crowdsourcing can improve innovation (e.g., collecting ideas, customer feedback, prototypes), marketing (e.g., using crowdsourcing as a market research tool, increasing brand image and customer loyalty as contributors feel their voice has been heard), after sales service (e.g., providing peer support and new service experiences) and HR processes (e.g., recruiting new employees, employer branding for digital natives). To fully capitalize on this multiplicity of business opportunities, crowdsourcers must develop success metrics that mirror all of them.

2. Start Small and Ensure Responsiveness

Initially, the crowdsourcing platform should be built around a small group of motivated employees who are convinced of the potential of crowdsourcing and are willing to actively work with the crowd. These employees must have access to sufficient resources to implement early contributions quickly. Expedient implementation demonstrates a crowdsourcer's willingness and ability to absorb crowdsourced data and satisfies the initial expectations of the crowd. As the crowdsourcer builds its absorptive capacity, the platform can expand to other domains and to more complex tasks.

3. Make Crowdsourcing "Cool"

Employees of crowdsourcers are much more likely to participate actively in crowdsourcing if they perceive it as a way of achieving personal successes and building their own reputation among colleagues. Thus, positioning the crowdsourcing platform as a vanguard project for further developing the company's way of working may help motivate employees to engage in crowdsourcing. This is particularly important in the platform's starting phase.

4. Post Precise and Understandable Tasks

The more precise and the more understandable the task is for the crowd, the less likely that crowdsourcing data will exhibit high variety. The key to success is to translate specific organizational problems into task descriptions that are clear, concise and self-explanatory.

5. Use Crowdsourcing for Experimentation

The rapid response from contributors enables crowdsourcers to receive feedback on tasks and posted questions almost instantly. This means that crowdsourcers can perform several iterations at very limited cost to support organizational learning. For instance, crowdsourcers can easily experiment with different task descriptions and formats to improve the quality of contributions. Similarly, employees can use crowdsourcing platforms as vehicles for testing, refining and prioritizing new ideas in multiple iterations without the need for expensive market research.

6. Involve the Crowd in Improving Data Quality

The huge volumes and high variety of data generated by crowdsourcing cannot be structured and filtered efficiently by employees. Thus, crowdsourcers should create collaboration-based processes in which contributors are integrated into tasks such as data structuring (e.g., adding tags, categories), filtering (e.g., identifying spam, duplicates), evaluating (e.g., voting, comments) and aggregating (e.g., rankings, trends). Contributors must be provided with incentives for engaging in collaboration-based crowdsourcing. Potential incentives could involve ranking and point systems, rights and role elevation or activity prizes.

Concluding Remarks

Crowdsourcing is a powerful approach for tapping into the collective intelligence of the broad-based community of Internet users. It can improve an organization's problem-solving capability, innovation, brand image, customer support and recruitment. This article provides recommendations on how organizations can build the absorptive capacity needed to capture this multi-faceted business value and to effectively overcome the challenges of implementing crowdsourcing and leveraging crowdsourced data.

Appendix: Research Methodology

As absorption challenges and capabilities of crowdsourcers are not well understood, we conducted multiple explorative and qualitative case studies[20] of four crowdsourcing platforms.

We selected the cases based on the size of the crowdsourcer and the crowdsourcing platform (i.e., number of contributors) to ensure we investigated a range of sizes. We focused on the software industry, where crowdsourcing is quite common (e.g., open source software). The high innovation rate of this industry makes absorptive capacity particularly important.

Between July 2009 and October 2011, we interviewed 14 key stakeholders (e.g., platform managers, R&D and marketing employees, and contributors), with each interview taking up to two hours. The interview guideline consisted of open questions on the crowdsourcing platform's vision and goals, the relationship of the interviewees with the crowdsourcer and their personal backgrounds, and the absorption challenges and capabilities. As absorption challenges and capabilities are complex constructs, we made use of the "critical incident technique,"[21] where interviewees were asked to describe situations in which they were part of data absorption or were able to directly observe these activities. We also reviewed internal documents and observed the crowdsourcing platforms for several hours a week and recorded our impressions. In addition, we created user accounts so we could observe the behavior of contributors and the crowdsourcer's employees for a period of up to 18 months.

We started our analysis by identifying the absorption challenges faced by crowdsourcers and the absorption capabilities being used to overcome these challenges.[22] Two of the researchers then identified superordinate themes for the crowdsourcers' activities in an inductive fashion. These themes were condensed to a first-coding scheme that was continually adapted to our data. Next, we constructed narratives for each case, detailing all information about the challenges and countermeasures carried out by the crowdsourcers. Finally, we analyzed the data from an absorptive-capacity perspective, which helped us to explain the

---

20 For a detailed discussion of the methodology employed, see Yin, R. K. *Case Study Research. Design and Methods,* Sage Publications, 2009.

21 The critical incident technique is an interview approach suited for investigating factors that strongly influence the success or failure of working processes. For more information, see Flanagan, J. C. "The Critical Incident Technique," *Psychological Bulletin* (51:4), 1954, pp. 327-358.

22 We thank Rayna Dimitrova, Andreas Haas, Vincent Kahl, Nadiem von Heydebran and Christine Wang for their support in data collection and analysis.

relationships between challenges and absorption capabilities.[23] The coding scheme was adapted to an absorptive-capacity perspective. The intercoder reliability of the final coding system was tested with 12 interviews. This gave a Cohen's Kappa of 0.71, which indicates good agreement.[24]

About the Authors

Ivo Blohm

Ivo Blohm (ivo.blohm@unisg.ch) is a post-doctoral researcher at the Institute of Information Management, University of St. Gallen, Switzerland. He holds a Ph.D. from Technische Universität München (TUM), Germany, where he graduated in Technology-Oriented Business Administration in 2009 (majoring in information systems, marketing and electrical engineering). His research interests include open innovation, absorptive capacity, crowdsourcing and business engineering. He runs various research projects that support customer-driven development of innovations and crowdsourcing.

Jan Marco Leimeister

Jan Marco Leimeister (janmarco.leimeister@unisg.ch or leimeister@uni-kassel.de) is a professor of Information Systems and has held the Chair for Information Systems at Kassel University since 2008. Since 2012, he has also been a professor at the Institute of Information Management, University of St. Gallen, Switzerland. He runs research groups on virtual communities, crowdsourcing, service management, collaboration engineering and ubiquitous/mobile computing, and manages several publicly and industry-funded research projects. His teaching and research areas include IT innovation management, service science, ubiquitous and mobile computing, collaboration engineering, e-health and business engineering.

Helmut Krcmar

Helmut Krcmar (krcmar@in.tum.de) is a professor of Information Systems and has held the Chair for Information Systems at the Department of Informatics, Technische Universität München (TUM), Germany, since 2002. He previously worked as Post-Doctoral Fellow at the IBM Los Angeles Scientific Centre, as Assistant Professor of Information Systems at the Leonard Stern School of Business, New York University, and at Baruch College, City University of New York. From 1987 to 2002, he was Chair for Information Systems, Hohenheim University, Stuttgart, Germany. His research interests include information and knowledge management, IT-enabled value webs, service management, computer supported cooperative work, and information systems in healthcare and e-government.

---

23 The theoretical framework underlying this analysis was presented at the Academy of Management Annual Meeting 2011 in St. Antonio, Texas.

24 Landis, J. R. and Koch, G. G. "The Measurement of Observer Agreement for Categorical Data," *Biometrics* (33:1),1977, pp. 159-174.

Data Monetization: Lessons from a Retailer's Journey

In today's era of big data, business intelligence and analytics, and cloud computing, the previously elusive goal of data monetization has become more achievable. Our analysis of a four-stage journey of a leading U.S. retailer identifies the potential benefits and drawbacks of data monetization. Based on this company's experiences, we provide lessons that can help other companies considering data monetization initiatives.[1,2]

Mohammad S. Najjar
University of Memphis (U.S.)

William J. Kettinger
University of Memphis (U.S.)

Data Monetization in the Supply Chain

Data is now being created and transferred at an unprecedented rate, fueling the growth in business intelligence and analytics (BI&A)[3] to discover opportunities for improving and innovating in supply chains and to enhance supply-chain collaboration.[4] In retailing, new supplier/customer ecosystems are emerging in which BI&A services are offered through a supplier portal, which can be cloud-based. Cloud-based BI&A platforms allow retailers and their suppliers to share data and analytics, often for a price. Or a company may monetize its data by exchanging it for other benefits (e.g., merchandising benefits). These data-sharing ecosystems often involve new players (e.g., public cloud platform providers and/or third-party data coordinators, negotiators or analysts).

Many companies would like to monetize their data. *Data monetization* is when the intangible value of data is converted into real value, usually by selling it. Data may also be monetized by

---

1 Cynthia Beath, Jeanne Ross and Barbara Wixom are the accepting senior editors for this article.

2 An earlier version of this article was presented at the pre-ICIS SIM/*MISQE* workshop in Orlando, Florida, in December 2012. We are grateful to Omar El Sawy and other participants at the workshop for their insightful comments. We would also like to thank the anonymous retailer and big data analytics company that provided so much time and insight concerning their experiences with monetizing big data.

3 For background information on big data and BI&A, see Chen, H., Chiang, R. H. L. and Storey, V. C. "Business Intelligence and Analytics," *MIS Quarterly* (36:4), 2012, pp. 1165-1188; Hopkins, M. S., LaValle, S., Lesser, E., Shockley, R. and Kruschwitz, N. "Big Data, Analytics and the Path from Insights to Value," *Sloan Management Review* (52:2), 2011, pp. 21-32; and Wixom, B. H., Watson, H. J. and Werner, T. "Developing an Enterprise Business Intelligence Capability: The Norfolk Southern Journey," *MIS Quarterly Executive* (10:2), 2011, pp. 61-71.

4 For a discussion on BI as an IT capability for supply-chain collaboration, see Rai, A., Im, G. and Hornyak, R. "How CIOs Can Align IT capabilities for Supply Chain Relationships," *MIS Quarterly Executive* (8:1), 2009, pp. 9-18.

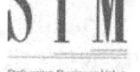

converting it into other tangible benefits (e.g., supplier funded advertising and discounts), or by avoiding costs (e.g., IT costs). Potential buyers of an organization's data include a direct supplier, an upstream supply-chain partner, a data aggregator, an analytics service provider or even a competitor. Three current IT trends are enhancing the potential for data monetization: big data, BI&A and the cloud.

Retail firms, with their exacting merchandising strategies and tight supply-chain relationships, have taken the lead in demonstrating that monetizing data can provide a significant revenue stream and be an IT cost-sharing mechanism. Point-of-sale, consumer-loyalty and inventory data can be sold to suppliers, and some of the cost of analyzing a retailer's data can be recovered from its suppliers.

Research has shown that data sharing in the supply chain improves supply-chain performance. Suppliers typically are interested in using a retailer's point-of-sale data to enhance planning and better manage inventory, thus reducing the bullwhip effect[5] (i.e., the phenomenon of demand variability amplification). Manufacturers can use downstream data about retail sales to improve product design, optimize operations and develop fact-based marketing and promotional campaigns. The availability of sales data to the supply chain means that demand can be more accurately forecast and, hence, inventory levels can be better predicted; in some cases, assemble-to-order can be achieved. Some suppliers may even use such data for strategic decisions by looking for product affinities to make merger or acquisition decisions.

Furthermore, data sharing can be a strategic tool in managing supply chains and channel relationships; sharing consumer or market data with supply-chain partners can influence their behavior.[6] Nevertheless, a company must decide

whether and when sharing its data with suppliers and other partners will pay off. The benefits a data-sharing strategy will have for the overall supply chain and distribution channel must be balanced against the benefits of holding data close to the chest.[7] While the improvement in supply-chain performance might be a good reason for companies to share data with supply-chain partners, a more explicit direct dollar value of the data can be another tempting motivation.

There are several challenges in involving suppliers in monetizing data. Selling data to suppliers may eliminate the competitive advantage that can be gained from asymmetric[8] information. Contracts have to be carefully prepared to ensure the data sold or shared is used for the mutual benefit of the firm and its partners. Trust has to be nurtured. The privacy and security of a company's data may be at risk if appropriate assurance practices are not established. Data packaging has to be considered to identify what data can be made available for sale and in what format and at what price. Pricing models need to be developed to take account of the associated cost of making data available and its value to the buyer. A company must identify a suitable marketing model for its data. Although data monetization best practices have yet to be identified, this article describes how a major U.S. retailer tackled these challenges. (The research we conducted to create this case study is described in the Appendix.)

Pathways to Data Monetization

Data monetization requires a strategic choice on which of several pathways to follow. It is important to assess the technical (data infrastructure) and analytical (human) capabilities of the company to determine which strategic pathway a company should choose for monetizing its data. The technical capability includes the hardware, software and network capabilities that enable the company to collect,

5   See Lee, H. L., Padmanabhan, V. and Whang, S. "The Bullwhip Effect in Supply Chains," *Sloan Management Review* (38:3), 1997, pp. 93-102.

6   For more discussion on the benefits of data sharing in the supply chain, see Zhou, H. and Benton Jr., W. C. "Supply Chain Practice and Information Sharing," *Journal of Operations Management* (25:6), 2007, pp. 1348-1365; Eyuboglu, N. and Atac, O. A. "Information Power: A Means for Increased Control in Channels of Distribution," *Psychology & Marketing* (8:3), 1991, pp. 197-213; Waller, M., Johnson, M. E. and Davis, T. "Vendor-Managed Inventory in the Retail Supply Chain," *Journal of Business Logistics* (20:1), 1999, pp. 183-203; and Lee, H. L., Padmanabhan, V. and Whang, S., op. cit., 2004, pp. 1875-1886.

7   For more discussion on the benefits of data sharing in the supply chain, see: Zhou, H. and Benton Jr., W. C., op. cit., 2007; Eyuboglu, N. and Atac, O. A., op. cit., 1991; Waller, M., Johnson, M. E., and Davis, T., op. cit., 1999; and Lee, H. L., Padmanabhan, V., and Whang, S., op. cit., 2004.

8   Information asymmetries occur when two people have different information about the same thing. If one has additional inside information, he or she can leverage or take advantage of that information.

Figure 1: Three Pathways to Data Monetization—Moving From Low-Low to High-High Capabilities

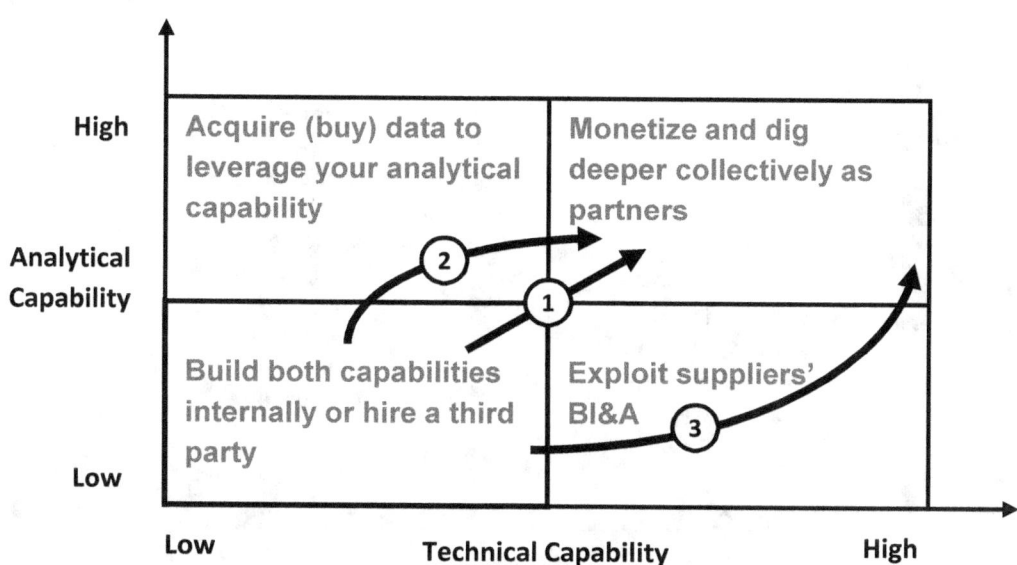

store and retrieve its data. The analytical capability is the mathematical and business analytical knowledge and skills of the employees in the company or in supplier firms. A company that has the data and the know-how (i.e., people and BI&A) to use the data properly will have an advantage in the era of big data. If both capabilities are low, then the company has three potential pathways to transition to the high capabilities that will enable it to monetize its data (see Figure 1).

Pathway 1: Move Direct to Higher Risk and High Reward

This direct pathway can be a riskier path to data monetization, as it requires simultaneously building both technical and analytical capabilities. As such, it requires the largest initial investment of the three alternative pathways. To follow this pathway, a company must invest in developing its technical infrastructure while hiring and training employees with the required business, mathematical and analytical skills. While costly, following this pathway will quickly position a company to be ready for monetizing its data and collaborating with supply-chain partners.

Pathway 2: Build Analytical Capability First

Following this pathway, a company chooses to develop its analytical capability first. This requires training employees and/or hiring business analysts with the required set of business, mathematical and analytical skills. As its analytical capability grow, the company may leverage them by generating more data (from internal sources) or buying data (from external sources). But growing an in-house analytical capability may not be sufficient to reach the point where the company can demonstrate the value of its big data and thus pave the way to data monetization. It may also require the company's technical capability to be expanded. This pathway requires a higher internal investment to develop the in-house analytical capability.

Pathway 3: Build Technical Data Infrastructure First

Instead of first developing its own analytical capability, a company may choose to extend or outsource its technical data infrastructure to produce an attractive collection of data that can be sold to suppliers. The creation of an appropriate digital platform is a prerequisite for a company and its suppliers to share data securely. A company can build this platform internally or use the expertise of a service provider; the use

Figure 2: DrugCo's Four-Stage Data Monetization Journey

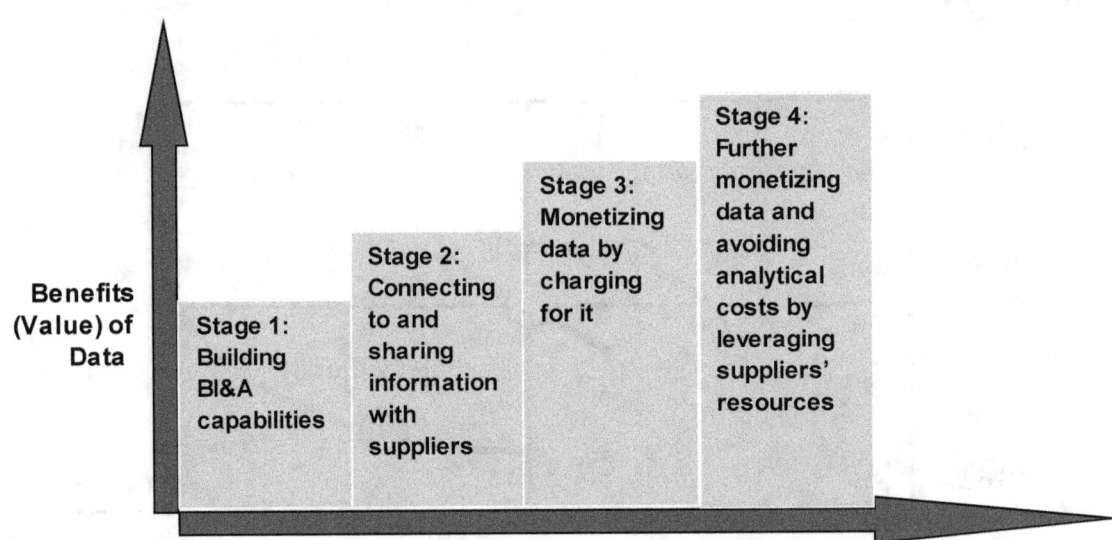

of cloud-based infrastructure can increase the flexibility, scalability and speed of developing the platform. By building a platform that will enable it to market its saleable data, a company can more quickly monetize its data and possibly avoid some analytical costs by leveraging the analytical capabilities of its suppliers rather than developing the analytical capability in-house. This pathway maximizes the potential data monetization pay-off because it enables sales of data and reduces startup costs. However, it does make the company more reliant on its partners as major sources of analytics.

The Data Monetization Journey of "DrugCo"

The case of "DrugCo," a U.S.-based Fortune 500 drug retailer with several thousand stores in more than half of U.S. states, illustrates a company that has followed Pathway 3. This company, which wishes to remain anonymous, is recognized as being relatively mature in BI and data use, and it has been monetizing its data for almost 10 years. The case shows how cost and the willingness to work with external parties and openly share data were important issues that motivated DrugCo to monetize its data.

Like other companies in the small-box retailing sector, DrugCo has:

- Many retail locations with narrowly defined geographical boundaries

- Limited shelf space

- Many stock-keeping units (SKUs) across the company

- A diverse customer base

- Differing inventories within each location to satisfy the local customer needs.

For DrugCo, data analysis is crucial for accurately assessing marketing campaigns, analyzing sales patterns, examining on-shelf availability and inventory levels, and customizing SKUs for each store based on its unique local consumer demand.

We describe key events that took place in the company, and we present a four-stage model that illustrates the four key stages it went through on its data monetization journey (the stages are depicted in Figure 2). We also provide lessons learned from DrugCo's journey for other managers as they grapple with their data monetization decisions.

In Stage 1, *Building BI&A capabilities,* DrugCo built its technical and analytical capabilities to address internal business needs.

In Stage 2, *Connecting to and sharing information with suppliers,* DrugCo connected to its supply-chain partners and started to share information with them through its cloud-based supplier portal, hosted by 3PP (a third-party data analytics firm that works with DrugCo and which also wishes to remain anonymous).

In Stage 3, *Monetizing data by charging for it*, DrugCo started selling its data to suppliers via its supplier portal.

In Stage 4, *Further monetizing data and avoiding analytical costs by leveraging suppliers' resources*, DrugCo leveraged its suppliers' data analytical capabilities and avoided some of the costs of its analytical function. This stage continues to the present day.

The characteristics of the four stages are described in Table 1. The stages differ in the technical and analytical (especially in people) capabilities the company required, the type of trust[9] built, the focus of DrugCo's information strategy, governance mechanisms, and the costs incurred and benefits achieved by various stakeholders. While there has been ample discussion of the first two stages, we were surprised by the third stage and even more surprised by the fourth.

As DrugCo moved from one stage to the next, the benefits realized from its data increased. DrugCo's data was monetized in the form of revenue generated directly from selling the data, as well as through a decrease in labor and infrastructure costs for analysis. The company also realized benefits from new business opportunities associated with new analytical insights and enhanced its collaboration with suppliers.

Stage 1: Building BI&A Capabilities

The growth of DrugCo's data sources meant that its traditional databases, database management systems and analytical tools became slow and inefficient. DrugCo's VP of Pharmacy Services described this environment:

> *"The database ... probably had about 1.2 to 1.3 million transactions a day and those transactions were very long ... there were literally hundreds of fields on one of these transactions that could be evaluated."*

In response, DrugCo improved its in-house technical data capability by developing a data warehouse and using basic data analytical tools

(e.g., Microsoft Access and Excel). Limited, functionally based BI capability was used to analyze and understand the implications of DrugCo's data. Business users would attempt to perform basic ad hoc queries and, when faced with more complex or time-consuming analyses, would ask the IT department for help. The main focus of this stage was to use data to meet business needs and solve internal problems. DrugCo's CIO described how limited capabilities meant limited analyses:

> *"If it takes you 45 minutes or an hour to get an answer... you're probably not going to do a lot with it. But if you can do it within 30 seconds or a minute or two, you are more likely to do more analytics and what-if cases."*

Because all data use was internal to DrugCo during Stage 1, inter-organizational trust was not an issue. Information was used to inform internal stakeholders and to run the business more efficiently. Data exploitation was judged to be going well since problems were being solved and new insights were being generated. Various policies were enforced to maintain the internal security and privacy of DrugCo's data.

The data exploitation costs in this stage were the technical cost of building the data warehouse and connecting it to the reporting tools, and the analytical cost of analyzing the data.

Stage 2: Connecting to and Sharing Information with Suppliers

In Stage 2, DrugCo created a secure, cloud-based portal for communicating with its suppliers. The portal provided access to point-of-sale, customer-loyalty and transactional data (e.g., purchases from DrugCo's suppliers) and various BI&A applications. As an analytical data warehouse platform, it allowed suppliers to work with and analyze DrugCo's data so the company and suppliers could collaborate on mutual business goals. DrugCo's Senior Director of Category Management Support (CMS) explained the importance of the supplier portal:

> *"The great thing about this portal and this information is [that DrugCo and its suppliers are] working on the same set of reports a lot of times and we're using the same information."*

---

9   Trust is categorized into contractual, goodwill and competence; see Sako, M. *Prices, Quality and Trust: Inter-firm Relations in Britain and Japan*, Cambridge University Press, Cambridge, 1992.

Table 1: Characteristics of the Four Stages of Data Monetization

	Stage 1: Building BI&A Capabilities	Stage 2: Connecting to and Sharing Information with Suppliers	Stage 3: Monetizing Data by Charging for It	Stage 4: Further Monetizing Data and Avoiding Analytical Costs by Leveraging Suppliers' Resources
Technical Capability	Implementing data warehouse with basic analytical tools	Developing a supplier portal	Extending the supplier portal with data integration and customized reporting capabilities for data	Offering a scalable data platform to accommodate expanded use of the suppliers' analytical capabilities
Analytical Capability	Internally focused, limited functional analytical capability	More fully developed internal and inter-organizational analytical capability	Matured internal and inter-organizational analytical capabilities; Learning what data is saleable	Exploiting analytical capabilities of suppliers
Type of Trust	Not an issue, as BI&A is internally focused	Contractual trust	Contractual trust; Goodwill trust	Contractual trust; Goodwill trust; Competence trust
Information Strategy	Informing internally	Supply-chain optimization	Revenue generation	Information transparency
Governance Mechanisms	Basic performance metrics; Information assurance	Information sharing contracts; Data presentation mechanisms and standards; Non-disclosure agreements (NDAs)	Pricing structure; Data purchase agreement; NDAs	Evaluation of supplier-provided analytics
Achieved Benefits/Associated Costs				
Achieved Benefits (DrugCo)	Data is used to meet specific business needs and solve problems	Data is shared across boundaries for supply-chain efficiency	Data is sold to generate monetary value and/or share technical costs	Data is traded for analytics to gain new insights; Cost savings and revenue growth
Associated Costs (DrugCo)	Technical cost; Analytical cost	Technical cost; Analytical cost; Contracting cost; 3PP's fee	Contracting cost; 3PP's fee	3PP's fee
Achieved Benefits (Suppliers)		Refined BI&A using the accessed data	Increased sales through better understanding of markets and DrugCo's business	Enhanced collaboration with DrugCo; Increased sales by shelf monitoring
Associated Costs (Suppliers)		Analytical cost; Contracting cost	Data cost; Analytical cost; Contracting cost	Data cost; Analytical cost

DrugCo owned the data it put on the supplier portal, while 3PP offered data-analytics, data-cleansing and consulting services, and owned the portal infrastructure. DrugCo sent its data to 3PP, which cleansed it and then uploaded it to the portal. Data security was enforced by preventing suppliers from copying or downloading data from the portal; they could only work with the data while it was still on the portal. Once it was connected with its suppliers, DrugCo had to further develop its analytical capability so it could respond to new inter-organizational analytical needs, which imposed additional analytical costs on DrugCo.

Trust is an important factor when external parties are involved with data monetization. In Stage 2, the data-sharing relationship between DrugCo and its suppliers was still somewhat immature. Non-disclosure agreements (NDAs) were used to specify what suppliers could and could not do with the data. These agreements created contractual trust—a mutual understanding between DrugCo and its suppliers based on the agreements. DrugCo's Senior Director of CMS described the contracting approach:

> "We've limited the use of the data. It's specifically limited to the purpose of growing the business of our company."

3PP acted as a liaison between DrugCo and its suppliers, providing value-adding activities by hosting DrugCo's data on the supplier portal, and BI&A services, administrating the information-sharing contracts, contracting directly with some suppliers (e.g., alcohol suppliers, which legally are not allowed to contract directly with DrugCo to purchase its data) and managing different aspects of the relationship, such as negotiating pricing of DrugCo's data.

During this stage, data was shared for supply-chain optimization. The suppliers accessed part of DrugCo's data, analyzed it and were able to enhance their marketing campaigns, production planning, pricing and inventory management.

The governance of DrugCo's supplier portal was designed to be collaborative. Major suppliers joined an advisory board that oversaw how the supplier portal was implemented. Voting was used to prioritize enhancements and to determine data presentation mechanisms and standards. The VP of Retail Solutions at 3PP explained the structure and function of the advisory board:

> "[At any time] there's around 18 to 20 suppliers on [DrugCo's] advisory board and there are eight that are on their senior council ... the larger group meets twice a year and the senior group meets four times a year ... they prioritize the changes or enhancements they want to see in the program and pass them to DrugCo ... DrugCo is only a member ... It's a user-driven advisory board."

DrugCo's costs during Stage 2 were the technical cost of building the supplier portal, the analytical cost for the additional inter-organization analyses, and the contracting cost for preparing contracts and NDAs with suppliers and third parties. 3PP incurred the cost of hosting the portal and providing additional analytical services. Suppliers connected to the portal also incurred contracting costs for the NDAs and analytical costs for analyzing the data they accessed. With direct access to the portal, suppliers could dynamically manipulate vast amounts of DrugCo data to answer questions on the fly.

Stage 2 laid the technical foundation (i.e., in the supplier portal) for data monetization and showed that DrugCo's data was valuable to its suppliers.

Stage 3: Monetizing Data by Charging for it

In Stage 3, with the supplier portal running successfully and suppliers having a good feel for DrugCo's data and its value, DrugCo began to extract more value from its data by monetizing it:

> "They [retailers in general] accumulate billions of records every year of point-of-sales transaction data and they are taking that huge amount of data and creating their own commercial data clouds for their suppliers to analyze ... A consumer-packaged-goods brand can just log in and see not only how their own products are doing in those stores but also how a competitor's products are doing in those stores." VP of Marketing, 3PP

The supplier portal was enhanced by adding additional data sets (particularly loyalty data)

Table 2: Four Levels of Data Packaging

Level	Data Access and Analytics Provided	Current No. of Suppliers	Percentage of Suppliers
Basic	• Supplier items only at POS transaction level detail filtered by SKU • Information provided shows supplier inventory level status • Access provided only through prebuilt reports	358	55.3%
Bronze	Basic Package plus: • Summaries for all approved classes/categories provided by a few prebuilt reports	128	19.8%
Silver	Bronze Package plus: • All items at POS transaction level detail for approved classes filtered by class • Ability to upload up to 10 GB of DrugCo's data for enhanced analysis by supplier • Third-party analysis tool provided for ad-hoc analysis by supplier • (Limited) basket view of categories a supplier operates in	82	12.7%
Gold	Silver Package plus: • (Full) basket view for all baskets, regardless of categories or supplier • Custom reports built for individual supplier or built for a set timeframe	79	12.2%

and customized reporting capabilities to provide a wider range of reports to the data-buying suppliers. DrugCo's internal and inter-organizational analytical capabilities matured, and it started to identify what data was saleable.

Data was offered in different packages, each of which had a different level of data granularity, reporting capability and price tag. By now, DrugCo had a dedicated executive on its merchandising team for selling its data, and this executive worked with 3PP to market these data packages directly to DrugCo's suppliers. Prices were often negotiated. If a supplier chose a higher level of information access and granularity, the price increased. There were four levels of data packaging—Basic, Bronze, Silver and Gold—for point-of-sale data (see Table 2). Only a limited number of DrugCo's major suppliers were allowed to purchase the highest Gold level package. As discussed later, a supplier had to invest resources in its relationship with DrugCo to become a candidate for the Gold level.

A data-purchase agreement and NDA were prepared for DrugCo and any supplier who wanted to buy data. Trust in Stage 3 included goodwill trust (based on beliefs) in addition to contractual trust (based on written agreements). When goodwill trust exists, partners are willing to go beyond stipulated contractual agreements. Thus, DrugCo trusted that the supplier would not only adhere to the data-purchase agreement, but would also use the data for the benefit of both parties. In essence, DrugCo's major suppliers learned to tell DrugCo when they saw a problem that needed to be addressed, regardless of whether doing so was of immediate benefit to the supplier.

Big suppliers (such Johnson & Johnson, Procter & Gamble, Coca Cola, PepsiCo, 3M, Novartis and Unilever) have been applying analytical tools for a long time to better predict demand and develop successful marketing campaigns; they are equipped with significant know-how in terms of BI&A:

"There are hundreds of CPG [consumer packaged goods] companies ... analyzing detailed data from retailers ... mixing it together with econometric and demographic data, weather data, various kinds of geographic data, and trying to better understand the markets and figure out how to better sell the products." Cofounder and CEO, 3PP

With access to more granular data, suppliers were able to fine-tune their operations by predicting sales trends more accurately and thus better develop marketing and promotional campaigns:

> "They [suppliers] can see a trial and repeat. They can see how a BOGO [Buy One Get One] type of promotional offer is performing, how our customers react to that differently than maybe a BOGO 50 [50% off] or a price point." Senior Director of CMS, DrugCo

During Stage 3, 3PP provided additional services to DrugCo, including training and supporting suppliers, negotiating and administering data-package contracts, BI&A services and marketing of DrugCo's data.

The information strategy of DrugCo at this stage shifted toward revenue generation; data was being sold and was generating a revenue stream for DrugCo. This revenue offset some of the costs of the underlying infrastructure, such as the data warehouse, the supplier portal and reporting tools.

Although DrugCo did not need to make additional investments in technical and analytical capabilities during Stage 3, it was still bearing 3PP's ongoing costs for hosting the cloud-based data and portal, and providing additional analytical services. It also incurred contracting costs for preparing the purchase agreements with data-buying suppliers. Suppliers were incurring the costs of buying DrugCo's data, negotiating the contracts for the data and analyzing the data. The suppliers benefitted by understanding the markets and DrugCo's business better. They were able to increase their sales by using DrugCo's granular data to design promotions and to leverage product affinities for additional promotional effectiveness. The Chairman & CEO of Procter & Gamble stressed the value of real-time, granular data:

> "For companies like ours who rely on external data partners, [getting the data] becomes part of the currency for the relationship. So as we deal with retailers, I may not be interested in getting that Tide ad this week, but if you give me your data in real time for the next four weeks,

that's more valuable to me ... It would be heretical in this company to say that data is more valuable than a brand, but it's the data sources that help create the brand and keep it dynamic."[10]

Stage 4: Further Monetizing Data and Avoiding Analytical Costs by Leveraging Suppliers' Resources

The final stage extended DrugCo's data monetization journey to new horizons, which enabled it to take even greater advantage of the analytical capabilities of its suppliers:

> "The purpose of that [suppliers having access to our data] is for them to be able to help us be smarter about how we run our business." CIO, DrugCo

The technical platform for DrugCo's data was expanded to meet new scale requirements arising from the suppliers' use of the platform to perform advanced analyses on the data. Also, advanced human capabilities were required to use applications that incorporated advanced analytical techniques (such as optimization, predictive modeling, simulation, time series modeling and principal component analysis). However, DrugCo avoided these additional analytical costs by exploiting its suppliers' analytical capabilities; it began to rely more on the business insights generated by suppliers' analyses of the data they purchased from DrugCo. The Cofounder and CEO of 3PP elaborated on the symbiotic relationship between retailers and CPG suppliers:

> "CPG companies are often quite sophisticated ... The retailers look at the CPG companies for advice [on] how to stock their shelves, how to do promotions, what products to sell, to whom [and] under what circumstances ... There's a symbiotic relationship in the sense that the retailer gets advice from the CPG company, and the more information the CPG company has about what's going on at the retailer and in the market, the better advice they would get, and of course there's the money angle

---

10 Interview with Robert McDonald, Chairman & CEO of Procter & Gamble, downloaded from http://www.mckinsey.com/insights/consumer_and_retail/inside_p_and_ampgs_digital_revolution.

... Retailers, like anyone else, are always looking for revenue sources and retail is a tough market, [with] very tight margins, and the more revenue they can get the better."

With access to DrugCo's data, suppliers started to understand the markets and DrugCo's business better; a supplier could get better insights into how it and DrugCo could together grow their businesses. This led, in turn, to DrugCo gaining a better understanding of its own promotions and its customers, and how they were buying products over time.

DrugCo's suppliers can now develop affinity analysis reports—which show what products are usually sold together—faster and more accurately, and pass these reports to DrugCo. The reports enable DrugCo to run separate promotions and advertising campaigns for highly related products instead of promoting and advertising them at the same time. The shift of data analytics to the suppliers resulted in a reduction of analytical costs for DrugCo.

Major suppliers offer insights to DrugCo through direct interaction on a daily basis between DrugCo's merchandising team and the suppliers' sales agents, often supported by BI&A analysts. In addition, supplier and DrugCo representatives are both involved in meetings of the supplier portal advisory board, where entire sessions may focus on analytics insights of benefit to DrugCo. For example, one major supplier presented a co-merchandising affinity-analysis program it had recently implemented, which predicts what third product will be purchased when two other products are bought. After reviewing the program, the advisory board voted and approved that it should be made available to Gold members, and it was included in the Gold level of data access.

In Stage 4, suppliers enhanced their collaboration with DrugCo and increased their sales; for example, they could use a shelf-monitor program that looks at sales of their products and detects a potential out-of-stock, which may cause a consumer to switch and buy a competitor's product. Some suppliers became trusted sources of data analysis. Based on these analyses, suppliers developed merchandising strategies and targeted promotional programs that DrugCo could implement:

"What we do with retailers [is] what we call Joint Business Planning or Joint Value Creation ... For us, getting data becomes a big part of value whereas for the retailer they have the data, so that's become a big part of our work together, and then how can we use this data to help them, because we have analytical capabilities that many retailers don't have, so often times we can use the data to help them decide how to merchandise or market their business in a positive way." Chairman & CEO, Procter & Gamble[11]

An additional form of trust, competence trust, was needed in Stage 4. DrugCo trusted that its partners had the superior managerial and technical capabilities needed to analyze its data. The company trusted that some suppliers had the capability and the willingness to use and analyze its data in a way that benefitted both parties while refraining from any misuse or misconduct regarding the data. 3PP's VP of Retail Solutions described how DrugCo's supplier portal enabled the formation of competence trust:

"[A retailer] would let their [suppliers] see the actual performance of the SKUs by day by store in a [market] basket level perspective because they were starting to trust the advice and counsel that their suppliers were giving them ... DrugCo can watch how the analysis was done by the [supplier] and argue it. The [supplier] really can't be sneaky because everything they do is wide open."

As DrugCo reached the fourth stage of the journey to data monetization, it shifted to a transparency strategy.[12] With this strategy, a company recognizes that the benefits of sharing data with external partners exceed those of withholding information from them. However, DrugCo realized the importance of limiting strategic information partnerships to the suppliers entitled to the highest Gold level

---

11 Interview with Robert McDonald, Chairman & CEO of Procter & Gamble, op. cit.

12 A transparency strategy is defined as one that selectively discloses information outside the boundaries of the firm to buyers, suppliers, competitors and other third parties like governments and local communities; see Granados, N. and Gupta, A. "Transparency Strategy: Competing with information in the Digital Age," *MIS Quarterly* (37:2), 2013, pp. 637-641.

data package. Allowing a supplier to purchase the Gold level package is viewed as a strategic merchandising decision and is based on the volume of transactions with the supplier, the number of people (i.e., the supplier's data analysts and salespeople) who are dedicated to work only with DrugCo and DrugCo's recognition of the supplier as a trusted advisor. Suppliers now compete to be designated by DrugCo as a "category captain." These suppliers review the performance of the entire category and recommend a store-level sales strategy, including assortment, shelf-space assignments, promotion, and pricing.[13] Category captains have the closest and most regular contact with DrugCo and invest time, effort and resources into the strategic development of their categories within DrugCo. They deploy dedicated analysts who only work with DrugCo and thus become trusted partners. In return, category captains have some degree of decision-making authority and an influential voice at DrugCo. DrugCo evaluates its suppliers' analytical performance based on the value of the analytics and recommendations provided by them and their track record of promoting DrugCo's business.

Lessons Learned

Several important lessons emerge from the DrugCo case. We believe the following practices will contribute to the successful monetization of data.

1. Consider How Creating and Sharing Data Will Change Relationships and Business Models

It is important to consider the dynamics among supply-chain members and to think about how data monetization might change the traditional relationships in the supply chain. Retailers can expect their major suppliers to compete for a category captain role to become a trusted advisor and a source of valuable business recommendations. Companies need to carefully consider the trade-off between higher levels of information transparency with their supply-chain partners and the possible risk of

losing information advantages over suppliers, customers and competitors.

Data monetization creates a new business model for the company, in which revenue generation, cost structure, value proposition and relationships change. The company's data is not only used to run the business, but also becomes a digital product the company can use to generate revenue and cover the costs associated with creating and gathering data. Leveraging suppliers' analytical capabilities introduces a new era of informational collaboration among partners and supply-chain members. Suppliers can add value to their relationships with retailers by offering business insights and new business-growth opportunities. Third parties can provide value-adding services to create and sustain a data monetization platform.

As the dynamics of competition and cooperation among companies continue to evolve, IT provides opportunities for value co-creation. A data monetization relationship is a good example of the co-creation of IT-based value between companies at the assets, complementary capabilities, knowledge-sharing and governance levels.[14]

2. Identify Where You Currently Are in the Data Monetization Journey and Where You Want to End Up

An ideal end state of a data monetization initiative will result in deeper insights from the associated ecosystem, a revenue stream, a reduction in infrastructure and analysis costs, and trusted use of data by supply-chain partners. The following are several aspects that concerned stakeholders have to pay attention to, prior to and during their data monetization journey.

Prepare Your Data for Sale. The integration of additional relevant data sets into the company's data will increase the value of the data to data buyers. For example, DrugCo enhanced the value of its data to its suppliers by adding loyalty data. Companies should also package the data for sale to meet different needs, analytical capabilities and willingness to pay. Multiple levels of data packaging (see Table 2) is a useful technique.

---

13   For an analysis and recommendations for choosing a category captain, see Subramanian, U., Raju, J. S., Dhar, S. K. and Wang, Y. "Competitive Consequences of Using a Category Captain," *Management Science* (56:10), 2010, pp. 1739-1765.

14   For more discussion on co-creating IT value, see Grover, V. and Kohli, R. "Cocreating IT Value: New Capabilities and Metrics for Multifirm Environments," *MIS Quarterly* (36:1), 2012, pp. 225-232.

Assess the Need for Value-Adding Third Parties to Join the Data Monetization Ecosystem. Third parties can provide various value-adding activities in the data monetization ecosystem. Examples include orchestrating the relationship between the company and the data buyer by hosting the data, contracting with data buyers, offering training and support, and providing technical and analytical capabilities. A third party can also be instrumental in the company's effort to obtain and build the required technical and analytical capabilities. Assessing what can be outsourced can be instrumental to building and sustaining a data monetization initiative.

Market Your Data and Challenge Your Suppliers to Get Onboard. A marketing strategy is needed to advertise and promote the value of the company's data. The company has to approach potential data buyers and highlight how and why the data is useful, as suggested by DrugCo's Senior Director of CMS:

> *"Challenge them saying: "Well, your competitors understand this better now. You know you're falling behind."*

Even when third parties participate in the data monetization initiative, the company still has to be involved in selling its data:

> *"You have to be involved with pushing it and selling it. You don't really outsource the selling of the data."* Senior Director of CMS, DrugCo

Avoid Some Analytical Costs by Leveraging Suppliers' Analytical Resources. A data monetization initiative can create new opportunities for the company to exploit its suppliers' ability to analyze data. It is not uncommon for there to be more analytical resources on the supplier's side dedicated to working on and analyzing the company's data, as highlighted by DrugCo's CIO:

> *"More [analytical] people on the [supplier] side have access to [our data] than we do internally."*

Recognize and Reward Your Top-Performing Suppliers. Determining appropriate measures to identify top-performing suppliers in your data monetization ecosystem and rewarding them will establish a collaborative relationship in which actions are guided by the principle of mutual benefit. A supplier can be rewarded by allowing it to have a higher level of data package and by nominating it as a category captain. Decisions to recognize top performance should not only be based on transaction volume, but also on the supplier's provision of human capabilities and the quality of advice provided. The performance of existing category captains should be continuously monitored so that underperforming category captains can be replaced with new ones.

3. Develop Contracts to Ensure Adherence to Data Monetization Policies

Several contracts were developed between DrugCo, 3PP and DrugCo's suppliers throughout the data monetization journey, notably NDAs and data-sharing and -purchase contracts. These contracts restricted the use of the shared or purchased data to specific purposes. Suppliers were obliged to use the data they purchased for the sole purpose of growing the mutual business of the suppliers and DrugCo.

4. Nurture Trust Between the Involved Parties

Different forms of inter-organizational trust exist between business partners. Trust can lower the contracting cost and conflict level required to reach a data-purchase agreement. The progression from trust based on written agreements to trust based on beliefs contributes to the formation of a collaborative relationship in which mutual benefits are considered by the parties involved. Inter-organizational trust can be built by communication of trustworthiness, inter-organizational coordination to establish governance mechanisms, and successful and repeated interactions that demonstrate each partner's reliability. The transparency of the collaboration portal can also nurture trust between a company and its suppliers; suppliers can be held accountable for their use of the company's data and the quality of the analysis and advice they provide.

Concluding Comments

The DrugCo case demonstrates that getting direct monetary value from a company's data is no longer elusive. Data analysis tools and cloud computing have paved the way to monetizing a company's data. We have described how DrugCo was able to monetize its data by going through four distinct stages and ultimately increased both tangible and intangible benefits. Building technical and analytical capabilities and connecting with the retailer's suppliers facilitated the emergence of a digital ecosystem that enabled data monetization. DrugCo managed to cut its analytical costs by leveraging its suppliers' well-established technical and analytical capabilities. Joint benefits emerged from this new relationship by generating a new revenue stream and providing a cost-sharing mechanism for the retailer, and offering suppliers real-time access to the retailer's data.

Appendix: Research Approach

The topic of data monetization arose when one of the researchers interacted with an executive of 3PP, a company that provides cloud-based big data hosting as well as analytical and consulting services. This firm had considerable experience with building supplier portals and/or cloud-based data ecosystems so companies could monetize their data. At the researcher's request, 3PP identified several of its clients that had monetized their data, and the researcher approached them about the possibility of in-depth cases concerning the "how and why" of data monetization. DrugCo was willing to discuss its journey on the condition that it remained anonymous.

First, we carried out numerous rounds of interviews at 3PP with the VP of Business Analytics, VP of Retail Solutions, Client Project Manager and Client Relationship Manager to more fully understand data monetization in general and 3PP's experiences with DrugCo in its role as a catalyst and facilitator of DrugCo's data monetization journey. The data provided by these interviews was analyzed and formed the initial picture of DrugCo's journey.

Next, data gathered from the interviews with 3PP was used to develop the interview guide to be used at DrugCo. Executives at DrugCo who were knowledgeable about and had participated in DrugCo's data monetization journey were identified with the help of 3PP. In-depth interviews were conducted with DrugCo's CIO, the Director of Category Management Services and the VP of Pharmacy, who provided details about DrugCo's journey. Email follow-up questioning also occurred.

Finally, follow-up corroborating interviews were conducted with 3PP's VP of Retail Solutions, Client Project Manager and Client Relationship Manager to triangulate accounts. Secondary sources, including some additional interviews at 3PP and public sources, complemented our primary sources and allowed us to form an overall view of data monetization.

About the Authors

Mohammad S. Najjar

Mohammad Najjar (msnajjar@memphis.edu) is a Ph.D. candidate at the Fogelman College of Business and Economics at the University of Memphis. He received his M.B.A. and B.Sc. from the University of Jordan. His research interests include IS services, business intelligence, information assurance and information management. He has published in and reviewed for several international conferences.

William J. Kettinger

William Kettinger (bill.kettinger@memphis.edu) is Professor and FedEx Endowed Chair in MIS at the Fogelman College of Business and Economics at the University of Memphis. Kettinger's focus is practical, rigorous research appearing in leading journals. He has received such honors as a Society of Information Management's Best Paper Award and directed a SIM APC study of the business drivers of IT value. He has served on the editorial boards of *MIS Quarterly, Information Systems Research, Journal of the Association of Information Systems* and *MIS Quarterly Executive.* He consults with global companies such as enterpriseIQ®, AT&T and IBM.

Exploiting Big Data from Mobile Device Sensor-Based Apps: Challenges and Benefits

Sensor data gathered from mobile devices generates big data, which can help organizations continuously monitor a wide range of processes in ways that prompt actions and generate value. However, generating reliable data from such devices is not straightforward and presents a variety of challenges. This article describes the challenges and provides guidelines based on a case study of "Street Bump," a mobile device app that the City of Boston uses to facilitate road infrastructure management.[1,2]

Daniel E. O'Leary
University of Southern
California (U.S.)

Sensor Data from Mobile Devices Provides New Opportunities

In the world of big data, organizations are increasingly turning to mobile devices as new sources of data derived from continuously monitoring a wide range of processes and situations. Mobile devices can facilitate the gathering of data from a diverse set of internal and external[3] stakeholders, including employees, customers and citizens willing to "donate" their data. Moreover, this data can be used in myriad ways. For example, sensor data on how an automobile is driven is providing big benefits for organizations ranging from government agencies to insurers.[4,5]

Facilitated by the "Internet of Things (IOT)," donated sensor data generated from users' mobile devices provides important new opportunities. As originally conceived, the IOT includes all kinds of devices connected to the Internet and to each other, generating sensor-based signals independent of human intervention. As noted by Ashton,[6] *"We need to empower computers with*

1   Cynthia Beath, Jeanne Ross and Barbara Wixom are the accepting senior editors for this article.
2   An earlier version of this article was presented at the pre-ICIS SIM/MISQE workshop in Orlando, Florida, in December 2012. In addition to the workshop participants, the author would like to thank anonymous referees and special issue editors, as well as Chris Osgood and James Solomon from the City of Boston.
3   Piccoli, G. and Pigni, F. "Harvesting External Data: The Potential of Digital Data Streams," *MIS Quarterly Executive* (12:1), 2013, pp. 53-64.
4   Vaia, G., Carmel, E., DeLone, W., Trautsch, H. and Menichetti, F. "Vehicle Telematics at an Italian Insurer: New Auto Insurance Products and a New Industry Ecosystem," *MIS Quarterly Executive* (11:3), 2012, pp. 113-125.
5   Mann, T. "Data Driven: New Program to Fix New York City Streets," *The Wall Street Journal,* August 30, 2013.
6   For a summary of the history of the Internet of Things, see Ashton, K. "That 'Internet of Things' Thing," *RFID Journal,* June 22, 2009 (http://www.rfidjournal.com/articles/view?4986).

their own means of gathering information, so they can see, hear and smell the world for themselves, in all its random glory ... sensor technology [that] enable[s] computers to observe, identify and understand the world—without the limitations of human-entered data." With this concept of the IOT, humans play a relatively passive role, simply ensuring that their mobile devices are switched on and generating data. As further noted by Ashton:[7] *"people have limited time, attention and accuracy—all of which means they are not very good at capturing data about things in the real world."* In addition, with activities such as driving, it is dangerous or illegal for people to simultaneously perform the activity and provide associated data. Mobile devices can be configured to capture that data and relieve humans of the responsibility.

Although sensor data from mobile devices can help generate deep insights and create value, gathering this kind of data does have its challenges. Using sensor data is not simply a matter of connecting devices together. Instead, companies need to identify and address the technical, behavioral, organizational and privacy issues arising from mobile device sensor initiatives.

Based on a case study of the City of Boston's "Street Bump" app available for iPhones, this article describes the emerging challenges and proposes approaches to overcoming them. With Street Bump, users place their iPhone in their cars, turn the app on and drive. The app captures data about the location of potential potholes that are encountered, which can then be uploaded to a central server. City of Boston managers actively mitigate the challenges arising from the app so that the data can be used to provide benefits both to the city and to app users.

The Street Bump App

The City of Boston has over 800 miles of roads and repairs over 19,000 potholes every year.[8] Historically, potholes have been identified by city workers. As noted by Mathew Mayrl from the City of Boston, *"We probably have 30 to 40 staff out there each day, and one of their responsibilities is*

to identify potholes."[9] In 2012, the City of Boston developed an app called "Street Bump" (see Figure 1) to allow virtually anyone to capture and report pothole data using their iPhone.

Figure 1: The Street Bump iPhone App

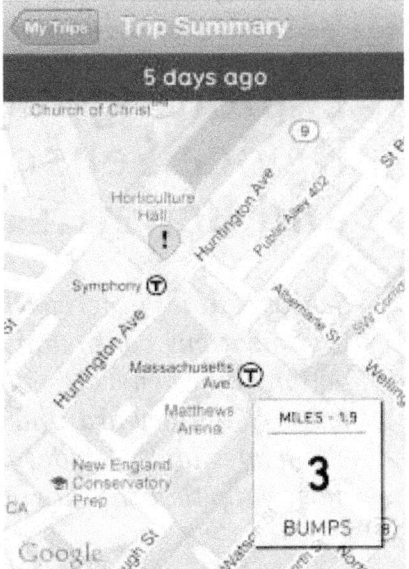

Source: http://www.newurbanmechanics.org/projects/streetscapes/bump/

In theory, drivers simply need to download and activate the app, put their phone on the dashboard or in a cup holder and drive. Using this approach, an army of mobile phone users driving the streets of Boston can replicate the work of city employees traveling block to block, surveying streets and looking for potholes. When a driver runs over a pothole, the phone captures the location of the pothole and then sends that location to a cloud-based computer, which records the reported pothole information with the location of other potholes so that they can be scheduled for repair in a timely manner. Specifically, as noted by Nigel Jacob from the City of Boston, *"[Street Bump] provides a new way for people to donate their data in solving public good problems."*[10]

Street Bump is being recognized as an important emerging innovation. For example, City of Boston Co-Chairs of the Mayor's Office of New Urban Mechanics (Nigel Jacob and Chris Osgood),

7 Ibid.

8 Ngowi, R. "App detects potholes, alerts Boston city officials," *USA Today*, July 20, 2012 (http://usatoday30.usatoday.com/tech/news/story/2012-07-20/pothole-app/56367586/1).

9 *Bump App Detects Potholes, Alerts City Officials,* Associated Press, July 20, 2012 (http://www.youtube.com/watch?v=yxAYLA405pU).

10 Ngowi, R., op. cit., 2012.

whose responsibilities include the Street Bump app, were awarded the 2013 Tribeca Disruptive Innovation Award.[11]

Street Bump Technology

The Street Bump app[12] uses the mobile phone's accelerometer to detect potential potholes, and the phone's global positioning system (GPS) capabilities to gather location information of that pothole. As users of the app drive around Boston, a constant stream of data is captured by the iPhones and analyzed for "bumps." Street Bump currently provides users with feedback in the form of a summary of the number of bumps and the length of the trip (see Figure 1). The user then decides whether to submit the bump data to the City Works Department, which is responsible for repairing potholes. If so, the data is uploaded to a server, and likely road problems are submitted to the city via open311,[13] where they are classified as potholes that need to be fixed or as known obstacles such as speed bumps.

GPS software maps the location of potential potholes with an accuracy of within 18 feet. If an adjacent bump is encountered, the two become a cluster, and the center is calculated as the midpoint between the two.

The cloud servers to which the app sends data are hosted by Connected Bits, Boston's partner on the project, with the data stored in a MongoDB database. City users can generate queries to analyze the data collected in that database.

Filtering Out Bumps that are not Potholes

Although Street Bump was designed for use by citizens, the initial implementation has been restricted to data gathered by city employees. In late summer and fall of 2012, 25 employees used the app to gather data on bumps in the highway. By the end of the year, the city realized that Street Bump was only about 10% effective in determining the existence of a genuine pothole, rather than a speed bump or a raised or damaged casting. Street Bump did not discriminate well between those different types of road anomalies.

---

11 http://www.tribecadisruptiveinnovationawards.com/?p=209.
12 The app can be downloaded at https://itunes.apple.com/us/app/street-bump/id528964742?mt=8.
13 Open311 is a collaborative model and open standard for civic issue tracking.

To improve the effectiveness, the city captured the location of known objects, which were bumps but not potholes. The app also uses different algorithms to filter out alternative sources of movement and bumps, such as joints in bridges, speed bumps, utility damage to streets and damaged street hardware (e.g., manhole covers).

These changes improved the pothole detection accuracy rate to between 30% and 40%. But even with this higher level of accuracy, managers still needed to have a person physically identify the existence of a pothole before it could be confirmed as one that needed to be repaired. Unfortunately, physical confirmation of the existence of potholes is a costly process; therefore, the city continuously works to improve the accuracy rate over time.

City officials also made the app widely available to the public to facilitate testing by individuals, which has resulted in useful feedback. For example, users discovered that the location of the phone in the car and the type of car could result in different data being generated for the same road anomaly.

Potential Benefits of Generating Data from Mobile Devices

Mobile device apps like Street Bump offer the opportunity to provide continuous monitoring and thus generate big data. For example, mobile phones constantly monitor location and other variables, such as heart beats, noise levels and movement. These capabilities can generate significant quantities of data and ultimately provide direct, indirect and emerging future benefits.

Direct Planned Benefits

Mobile apps can continuously monitor situations and conditions from many different sources or locations and from a wide range of users. The cost of gathering this data generated in real time is potentially low. In addition, it can be analyzed and integrated with business processes to solve problems.

The cost of finding and fixing potholes in the United States is substantial. For example, in the winter and spring time, New York City has 40 crews working 24 hours a day, seven days a week, simply finding and filling potholes.[14] In the case of Street Bump, rather than having crews find the potholes, that task can be carried out by mobile phone owners, thus reducing personnel, equipment and other costs for the city.

There are many individual stories of how potholes have "destroyed" everything from school buses to Ferraris. Reportedly, deaths have even resulted from potholes.[15] Street Bump allows both major anomalies in the road infrastructure and "emerging" potholes to be identified so that they can be fixed before they become a major traffic hazard, thus reducing the costs of potholes to the public. Speed of identification is particularly important because the cost of fixing potholes can increase exponentially as the quality of the road deteriorates.

Indirect Benefits

As data is gathered using mobile apps, managers gain real-time insights into the processes being monitored. Continuous monitoring generates data about the process, facilitating transparency and eliminating information asymmetries.[16] As data is gathered using mobile apps, managers can align their actions, strategies, processes and organization with the information and knowledge generated from the processes being monitored.

Organizations can use the insights into the information asymmetries to generate competitive or political advantages over competitors so they win the support of customers, voters, etc.

New data also provides the opportunity to change processes and organizational relationships. When it implemented Street Bump, the City of Boston found that it had over 300,000 castings on the city streets and sidewalks. Although some belong to the city, the majority belong to other organizations such as the water company. This has led to monthly meetings to keep both sides informed and allows tracking and management of the condition of those castings.

---

14 "City Grapples with Thousands of Potholes," *NY1 News*, February 9, 2011 (http://www.ny1.com/content/top_stories/133598/city-grapples-with-thousands-of-potholes).

15 Reeves, J. "Death by pothole," *The Post and Courier*, December 22, 2010 (http://www.postandcourier.com/article/20101222/PC1602/312229952).

16 Information asymmetries occur when two people have different information about the same thing. If one has additional inside information, he or she can leverage or take advantage of that information.

Potential Future Benefits

In the case of Street Bump, the plan is to integrate pothole identification capabilities into other systems that can facilitate the pothole repair process. For example, Street Bump could be integrated with a system that allows rapid response and resource allocation in the form of a pothole management system. Such a system could provide an improved capability to facilitate management of city resources and help guide resource allocation decisions and infrastructure investment.

Street Bump has additional potential future benefits. For example, the existence of potholes may indicate other highway management problems, such as bridge joint disrepair or manhole cover disrepair or of the need for more extensive road repairs. Pothole data may also support the analysis of other city services. For example, to what extent are potholes predictive of problems leading to police and fire calls? To what extent do potholes limit the speed of response of the police and fire services? Linking real-time quality pothole databases to other databases enables these and other related problems to be investigated.

Challenges of Sensor-Based Mobile Device Apps

Although sensor-based mobile device apps can provide significant benefits, there are also challenges. We illustrate some of the challenges by describing the experiences with Street Bump.

Information Systems Management Issues

Sensor-based mobile device apps present challenges that require management decisions. Managers need to consider what types of devices an app will be available for and the scalability of the supporting infrastructure. They also need to ensure that the data collected is of sufficient quality to meet business requirements.

Device Standardization. Sensor signals can vary according to the particular type of mobile device, and various device characteristics, such as the hardware and operating system, can impact the signal. For example, the accelerometers used in different mobile phones vary. Thus, given the same input, the movement and shocks captured by different types of mobile phones can be different. Accordingly, to generate a consistent sensor signal, it may be necessary to standardize on a single type of device, at least initially. Alternatively, additional measures must be put in place to make the signals and their interpretations similar.

In the case of Street Bump, the iPhone accelerometer, the Android accelerometer and other phone accelerometers and their operating systems generated different signals from the same bump in the road. Further, different Android phones had different accelerometers. As a result, in order to ensure signal reliability and data consistency, the City of Boston decided that, initially, data would only be gathered from iPhones.

Infrastructure Scalability. Scalability of the supporting IT infrastructure can be a key concern when collecting data from mobile devices. As use of an app moves from a few users to many users, facilities can be overwhelmed by increasing data volumes.

In the case of Street Bump, the City of Boston mitigated some of the scalability issues by initially capturing Street Bump pothole information only from City employees. This prototyping approach was taken so that officials could begin to understand some of the key data management concerns associated with data volumes before having to process and store the data from a larger number of users.

Data Quality. The quality of technology in mobile devices can limit the quality of the information gathered from the devices, compromising the precision and limiting the use of the data for its intended purposes. The sensitivity of components such as accelerometers or GPS capabilities differs according to the device.

With Street Bump, location information from iPhones is accurate to within 18 feet; therefore, multiple sources of the same anomaly can be used to triangulate to get a better fix on the actual location. For example, when multiple users find an anomaly at the same or a similar location, the redundancy in the data generated provides better evidence for the existence and location of a potential pothole or cluster of potholes.[17]

Data Donation Management Issues

Controls need to be considered when mobile devices are used to collect data donations from the public. Organizations must also plan how they will respond to the donations. Accordingly, managers must consider the context of the device use, user behavior and user incentives and expectations.

Context of Device Use. Although gathering sensor data from mobile devices may seem relatively mechanistic, such data can be influenced by the use of the device, including the context in which it is used. For example, Street Bump users found that the location of the mobile phone in the car, the type of car and which phone is used make a difference to whether or not a pothole is identified. One user noted that he drove the same route with two different cars. The first car, a four wheel drive vehicle, found 50 different potholes, whereas the second, a Jaguar, found only one over the same 3.5 mile journey. That same user noted that the location of the phone in the car also made a difference on whether potholes would be detected.

User Behavior. User behavior can also influence what the device finds. For example, in the case of Street Bump, a driver might consciously slow down or drive around potholes to better preserve his or her car, which clearly limits the potential success of the app in finding potholes. Alternatively, although unlikely, drivers, intent on providing the city with detailed pothole information could speed-up as they drive through potholes. Each behavior would result in different findings from the exact same environment.

User Incentives and Expectations Management. A key challenge in using a mobile app to facilitate data collection is to ensure the app is used. If the app is too costly or there are no incentives for using it, there are not likely to be many data donations. In the case of Street Bump, the app was available as a free download, minimizing the direct costs to iPhone users. Further, users likely anticipated that by driving with the app on in their car, potholes would be identified and the city would use that information to fix the potholes. There would be a direct user benefit because potholes on the roads they used would more likely be identified and fixed. In addition, users may have "felt good" about being part of a network that provided signals to

---

17 "Boston releases Street Bump app that automatically detects potholes while driving," *MailOnline*, July 20, 2012 (http://www. dailymail.co.uk/news/article-2176783/City-releases-motion-detecting-Street-Bump-app-automatically-detects-reports-potholes-driving.html).

monitor road quality. All in all, there would be definite incentives to using Street Bump.

When users donate data, they likely have expectations for how that data will be used. To accommodate those expectations, organizations need to provide feedback in different situations. However, once the signal is received by the organization, what kind of feedback should occur, and what events should drive the feedback to the information provider? To complicate the issue, since mobile devices work passively while users are performing other activities, it may be that users do not notice that the devices are generating usable data. As currently configured, Street Bump does not capture information about users, which means there is no mechanism for providing feedback on whether a pothole was fixed, thus limiting the ability to link expectations and incentives.

The sensor data generated by Street Bump can influence expectations. If Street Bump finds a bump, citizen expectations are that the city should be able to fix it. However, Street Bump identifies many more casting bumps than actual potholes, but most of those castings are not owned or maintained by the city.[18] As a result, the city cannot fix many of the bumps identified by Street Bump.

Application Management Issues

Application management is concerned with delivering the expertise, skills and solutions necessary to successfully build, run and evolve applications. Applications provide the interface between the IT infrastructure and the user. Designing user-friendly interfaces requires the appropriate domain expertise. In addition, organizations need to understand the issues associated with the inherent biases of using mobile devices.

Domain Expertise. Organizations may or may not have the necessary expertise to develop the specific app. Those without the appropriate expertise must consider alternative options such as outsourcing. The City of Boston outsourced the development of Street Bump, removing the need to acquire internal expertise about the app. The outsourcer was chosen as a result of a "challenge"

(or competition) that invited interested parties to propose solutions.[19]

Inherent Biases. Using mobile devices for data collection can result in various inherent biases. To participate, a user must have access to the device and app, and the context. If users do not have access to any of those three, they will not be involved in generating data, and resource allocation decisions may be made to their detriment. As a result, using sensor-generated data can result in a bias—a kind of "digital divide"[20] between participants who can afford the app, device and context, and non-participants who cannot afford them.

To capture road infrastructure information with Street Bump, the user needs to have the app, an iPhone and a car. Although the app is free, not everyone has access to an iPhone and car. This may result in disproportionate amounts of data being generated about some roads, which can result in greater attention being given to road infrastructure in some locations. To date, the City of Boston has circumvented this issue because only city personnel are used for data capture.

Privacy Management Issues

Mobile apps provide the ability to capture substantial broad-based sensor information generated by individuals based on their behavior. Although users can limit which information they will share by turning the app on or off, potentially confidential information could be shared inadvertently, or life-style information could be embedded in the donated data. As a result, data donation or data sharing can result in both the perception of and actual invasion of privacy. Privacy concerns may also cause the user to be selective about when the app is switched on. In addition, organizations may have to forego the advantages of personalization and contextualization because of privacy concerns.

Perceived Privacy Invasion. The perception of an invasion of privacy is illustrated by accounts of how cities' use of other technologies has compromised privacy. For example, there have been reports on how traffic cameras have captured cheating spouses running red lights; when the pictures were sent to their home as evidence, their spouse saw them with someone

18   Moskowitz, E. "App shows jarring role of cast-metal covers in Boston," *The Boston Globe*, December 16, 2012 (http://www.bostonglobe.com/metro/2012/12/16/pothole/2iNCJ05M15vmr4aGHACNgP/story.html).

19   Osgood, C. "Challenge: Build an App to Detect Potholes," (http://www.newurbanmechanics.org/2011/02/09/challenge-build-an-app-to-detect-potholes/).

20   Norris, P. *Digital Divide*, Cambridge University Press, 2001.

else. Thus, an important question to address is to what extent users will perceive sensor-based mobile apps that can track their every move as an invasion of their privacy?

Actual Invasion of Privacy. There can be privacy issues associated with knowledge about the user's life style that can be discovered from the donated data.[21] Since sensor data from mobile devices is gathered unobtrusively, and potentially continuously, data could be gathered that would be intrusive to the data provider's life. For example, the sequence of potholes provided by Street Bump could result in information about a route to a hospital or doctor. If that information were made available to others, it might be used against the data provider in situations such as insurance policies.

The City of Boston aimed to build a "strong relationship of trust" as part of its citizen-centric approach. As a result, when capturing bump location, any user information is stripped from the data. By stripping out user information, it is impossible to generate "user profiles" to analyze or investigate particular user behavior, thus assuring data-provider privacy.

Addressing privacy concerns can result in substantial costs. If data is cleansed of many of its identifying features, the ability of the organization to generate knowledge from data can be limited. In the case of Street Bump, eliminating user and pothole sequence information can severely limit additional knowledge that might be mined from the data.

Selective Use to Preserve Privacy. As a result of privacy concerns, users are likely to be selective about the information that they disclose to the system. In particular, privacy issues may influence the time when a user turns on a sensor-based app and how long it is turned on for. In the case of Street Bump, users may be selective about the routes for which they allow the system to capture data. For example, the user might turn the app on for day-to-day activities, such as driving to work, but not turn it on for after-work activities or trips for medical purposes. This behavior can lead to potential biases in the information that is generated.

Personalization and Contextualization Privacy Issues. Different users have different preferences. As a result, information is often gathered from users to personalize their usage. Unfortunately, the more personalization built into the mobile device and app, the greater the potential loss of privacy. For this reason, the City of Boston has chosen to exclude all personal information from Street Bump's use to provide maximum privacy.

Similarly, organizations can choose the extent to which device capabilities are "optimized" for the specific context. In particular, organizations determine the extent to which they will account for context variables that influence mobile device sensor signals. In the case of Street Bump, the potential context information includes the type of device (iPhone), relevant technology within the device (GPS, accelerometer), the location of the device (e.g., car type and location of device within the car) and other factors.

The more fully the context information is considered and used, the "better" the potential performance of the app at capturing sensor data to accomplish the task at hand. Unfortunately, in some cases, providing context information can result in loss of privacy. For example, if the user employs a unique context or the context includes user characteristics, then the context can point directly to a particular user. Thus, in the case of Street Bump, there currently are no plans to gather context information from citizens since the city wants to preserve privacy.

Recent and Future Street Bump Developments

Street Bump is still evolving. Recent developments over the last year, some of which have been mentioned above, include:

- At the time of writing, the donated data is generated only by city employees. However, the City of Boston expects that as the accuracy of pothole detection increases, it will begin using data donated by citizens.

- To date, the city has not actively tried to engage citizens in data collection since it is still piloting Street Bump. As a result, less than 5,000 users worldwide have downloaded the app. Although data from users in other cities and countries is captured, it is neither mapped nor used.

---

21 O'Leary, D. "Knowledge discovery as a threat to database security," *Knowledge Discovery in Databases*, AAAI Press/MIT Press, 1991, pp. 507-516.

- Because of recent national and international concerns with privacy,[22] Boston has discussed the potential development of a "nutritional label" that would indicate the key privacy protection built into the app.

- The city is considering developing an app that is similar to Street Bump but is only used by city employees. The revised version would allow capture of specific trip information, thus allowing more detailed data capture than would be possible with data donated from the public.

- The city expects that, in the future, Street Bump will become an open source app, thus facilitating collaboration with other cities and developers.

Guidelines for Avoiding the Pitfalls of Sensor Data Generation

Based on the City of Boston's experience with Street Bump, the following seven guidelines will help organizations avoid the pitfalls of sensor data generation.

1. Make Data Actionable

For management to depend on and use the data generated from mobile device sensors, it must be timely and reliable—i.e., it must be actionable. In general, such data is timely but may not be reliable. In the early stages of a project, it is important to assess the reliability of the data being captured, and then work over time to improve reliability to achieve an acceptable level for the data's ultimate intended use.

2. Prepare for Likely Organization and Process Changes

The rate at which mobile devices and apps collect sensor data may outstrip the organization's ability to react to the insights generated from that data. For example, StreetBump can find potential potholes faster than the city can fix them. To reconcile this disconnect, managers may need to reengineer processes within the organization—for example, by leveraging automation or alerts—so that the organization can process and act on sensor data appropriately.

3. Design Personalization and Contextualization to Match Privacy Requirements

Organizations should determine the extent to which users can personalize apps and design them accordingly. Whereas personalization enables users to better meet their needs, the greater the level of personalization, the greater the potential loss of privacy. These conflicting influences must be balanced by managers who understand the unique risks of the initiative.

4. Plan for Growth

Organizations should anticipate the potential growth of sensor data and leverage advanced technologies, such as cloud computing and MapReduce,[23] to facilitate scalability. Sensor data can be divided into multiple pieces—either at the device level or event level—so that the data can be processed in parallel, which further facilitates scalable solutions.

5. Set a Narrow Band of Device Standards and Expand over Time

Organizations should initially choose to standardize on a particular device to facilitate sensor signal standardization from mobile devices. Over time, the device options can be extended to include related devices (e.g., iPads in the case of the iPhone) or emerging new devices that gain in popularity (e.g., in a recent poll, *USA Today* readers felt that Android users would outnumber iPhone users in three years.[24])

6. Pilot Test with Employees

When possible, organizations should pilot test mobile device sensor apps internally to more fully understand their use and effects—and to mitigate risks associated with unexpected issues from engaging with customers or external stakeholders. Pilot testing will show how user behavior with the mobile devices can influence

---

22   Satran, R. "NSA Data Fight Could Signal Privacy War," *US News and World Report*, July 30, 2013 (http://money.usnews.com/money/personal-finance/mutual-funds/articles/2013/07/30/nsa-data-fight-could-signal-privacy-war).

23   Dean, J. and Ghemawat, S. "MapReduce: Simplified Data Processing on Large Clusters," *Communications of the ACM* (51:1), January 2008, pp. 107-113.

24   *USA Today*, June 26, 2013, p. 1.

results. For example, in the case of Street Bump, snow and ice will likely cause drivers to slow down and be more cautious, which will affect the sensor data recorded for potholes.

7. Offer Incentives and Feedback while Managing Expectations

A mobile sensor app ultimately needs to make a difference in the life of the data provider, so organizations need to ensure that users can identify the value proposition associated with donating data from their mobile apps. Organizations should continually gather information about user costs and benefits, possibly by running experiments or by conducting periodic surveys.

Concluding Comments

The huge amount of data generated from the sensors in mobile devices can offer unique opportunities to provide continuous monitoring of a range of processes, in real time, potentially mitigating information asymmetries. But capturing and making use of the sensor data is not as simple as just adding a new data source. This article has described the challenges organizations face with sensor-based mobile device apps and provided guidelines for avoiding the pitfalls.

Appendix: Research Methodology

This study is based on data collected between June 2012 and July 2013. Information was gathered from multiple sources, including:

- *Social media:* Street Bump and Connected Bits have Twitter accounts that provide insights.

- *News media:* Several videos (for example, by the Associated Press) about Street Bump are available on YouTube. These videos include interviews with City of Boston personnel. In addition, news media, ranging from *USA Today* to *The Boston Globe,* have published articles about Street Bump.

- *Google Group:* There is a Google Group devoted to Street Bump. As part of the research, the author monitored contributions to that group.

- *Phone interviews with City of Boston personnel:* The author interviewed James Solomon (Street Bump Manager, Mayor's Office of New Urban Mechanics) and Chris Osgood (Co-Chair, Mayor's Office of New Urban Mechanics), both from the Public Works Department. Without their help, this article would not have been possible.

About the Author

Daniel E. O'Leary

Daniel O'Leary (oleary@usc.edu) is a professor in the Marshall School of Business at the University of Southern California. He is a former editor of *IEEE Intelligent Systems* and the current editor of John Wiley's *Intelligent Systems in Accounting, Finance and Management.* O'Leary is the author of *Enterprise Resource Planning Systems,* published by Cambridge University Press, which has been translated into Chinese and Russian. His current research focuses on technology-based crowdsourcing and other emerging technologies.

A Cubic Framework for the Chief Data Officer: Succeeding in a World of Big Data

A new breed of executive, the chief data officer (CDO), is emerging as a key leader in the organization. We provide a three-dimensional cubic framework that describes the role of the CDO. The three dimensions are: (1) Collaboration Direction (inwards vs. outwards), (2) Data Space (traditional data vs. big data) and (3) Value Impact (service vs. strategy). We illustrate the framework with examples from early adopters of the CDO role and provide recommendations to help organizations assess and strategize the establishment of their own CDOs.[1]

Yang Lee
Northeastern University (U.S.)

Stuart Madnick
Massachusetts Institute of Technology (U.S.)

Richard Wang
Massachusetts Institute of Technology (U.S.)

Forea Wang
Stanford University (U.S.)

Hongyun Zhang
Xi'an Jiaotong University (China)

The Need for a Chief Data Officer

Increasingly, companies expect that big data, with its focus on volume, velocity, variety, veracity, and value,[2] will be a powerful strategic resource for uncovering unforeseen patterns and developing sharper insights about customers, businesses, markets and environments. For example, some hospitals are applying automated learning algorithms to patient data and insurance claims data to discover new patterns and insights. The text in mountains of patient satisfaction survey data and data from social media, a kind of unstructured big data, can now be mined to analyze patients' sentiments about a hospital. As a result, U.S. hospitals can now determine how to improve their patient satisfaction scores, which are directly tied to the federal government's reimbursements to the hospitals.

Organizations need to determine who should manage big data. Data scientist roles have emerged to capitalize on the analytical opportunities of big data, but placing these specialists

---

1 Jeanne Ross, Cynthia Beath and Barbara Wixom are the accepting senior editors for this article.

2 Almost everyone agrees that big data is important, but few can agree on a definition. In our research, we define big data as having five key characteristics, known as "the five Vs": volume, velocity [the speed at which data is created], variety, veracity, and value. Note that veracity and value are often omitted from big data definitions. Veracity is concerned with data quality; high volume and high velocity is of limited use if the data is of poor or uncertain quality.

in operational business units without leadership at the corporate level might be insufficient to harness the full value of big data. A survey[3] of nearly 600 global executives revealed that most companies are still learning how to manage big data at the enterprise level. The survey also revealed that companies with a top executive responsible for data management have better financial performance than their peers.

To address the challenges and opportunities of big data, leading organizations have established a new breed of executive, the chief data officer (CDO). Wikipedia describes the CDO role as including "... defining strategic priorities for the company in the area of data systems, identifying new business opportunities pertaining to data, optimizing revenue generation through data, and generally representing data as a strategic business asset at the executive table."[4] In reality, although some CDOs strive to exploit big data for business strategy, others focus solely on data preparation for external reports, overseeing compliance and establishing data governance.

Emergence of Chief Data Officers

Leading organizations have learned an important lesson—that seemingly tedious data problems are often fundamentally business problems. As such, data problems can reflect weaknesses in business strategy and operations. Traditionally, organizations have addressed data problems by assigning a small group within the IT department to clean up data. As it has become evident that data problems, particularly business problems rooted in data problems, cannot be solved by the IT group alone, organizations have appointed data managers with a variety of titles, such as data quality managers, data quality analysts and data stewards. Data-governance mechanisms, committees, councils and workgroups have also been developed to identify and solve data-related problems and resolve conflicts. Finally, enterprise architecture and data

architecture have also been employed to align data, IT, and business processes and strategies.

Despite these efforts, organizations have continued to face data issues, and their ongoing concerns have led a growing number to establish an enterprise-level, executive-rank CDO. Some might argue that traditional data-related managers and data-governance mechanisms can deliver the same results as a CDO. However, there are critical differences between the efforts of low-level data managers and those of executive-rank CDOs. The key contrast lies in organizationally sanctioned leadership and the accountability given to executive-level CDOs.

First, unlike data managers, a CDO can lead the effort to build organizational capability that can energize and sustain the entire organization and extended enterprise. The experience of a major U.S. healthcare institution illustrates the inherent challenges faced by data managers who lack the authority of a CDO. While attempting to re-examine the business processes that collect, store and use customer data, a data quality manager in this institution received this complaint from an executive: *"You are digging in my backyard— Get out of my backyard!"* Another data manager recalled the project as: *"A huge responsibility without authority."* As a result of these obstacles, the entire project was discontinued; the group working on the project was dismantled and some members left the company. In reality, low-level data managers are not in a position to dictate business process changes to higher rank executives, let alone external partners.

The second critical difference between a CDO and traditional data managers or data-governance mechanisms is that the CDO can be held accountable for a failure of leadership in resolving data problems. Data-governance mechanisms, such as data-quality and -governance councils, committees and workgroups, can be useful for continuous improvement of data policies or standards, conflict resolution, and for reconciling and authorizing data sources. However, because individuals have responsibilities outside of the committee or workgroup, they are usually not held accountable for governance results.

Note that the CDO does not replace the need for data managers or data governance. Rather, the CDO leads data managers and enhances the

---

3 A summary of current global big data practices can be found in "Big data: Harnessing a game-changing asset," *Economist Intelligence Unit*, June, 2011. It explains in detail the current practices of 586 global companies with easy-to-understand tables and figures.

4 The Wikipedia page on the chief data officer (http://en.wikipedia.org/wiki/Chief_data_officer) is a good starting point for scanning the industry's current trends, but it does not provide a complete picture.

effectiveness of existing governance by putting data on the organization's business agenda and in the minds of other executives and officers. Under the leadership of a CDO, business strategies reflect and exploit data, particularly big data, instead of treating data merely as a by-product of running the business.[5]

The History of the CDO

The first recognized CDO was established in 2003 at Capital One. Yahoo! and Microsoft Germany were also early adopters of the CDO role. More recently, CDOs have been established at global investment banks, consumer banks, consumer credit institutions, financial institutions, IT and data companies, healthcare organizations, U.S. federal and state governments, and U.S. military organizations. For example, the U.S. Federal Communications Commission (FCC) created in each of its Bureaus a CDO with varying rank and scope; in total, the FCC created 11 CDOs. According to GoldenSource's annual client survey, "over 60% of firms surveyed are actively working toward creating specialized data stewards, and eventually chief data officers."[6]

Many organizations recognize that they need an executive to lead data management, but not necessarily with a CDO title. These full-time CDO-equivalent executives lead enterprise-wide initiatives on data quality and analytics, data governance, data architecture and data strategy. In this article, we use the term "CDO" to refer to all executives who are carrying out enterprise-level CDO roles, even if they may not formally be titled as CDOs.

CDO Reporting Relationships

As organizations use more advanced business analytics, often there is a need to redirect the flow of information horizontally across the enterprise. Thus, many of the CDOs and executives we interviewed[7] had the power to exert influence on company strategy. This power and authority is often reflected in their reporting relationships, membership on senior management committees, and authority over budgets and employment.

Of the CDOs we interviewed in our study:

- 30% reported directly to CEOs
- 20% to COOs
- 18% to CFOs.

Others reported to the CIO, CTO, CMO (chief medical officer) or CRMO (chief risk management officer). Many CDOs are members of senior management committees and have the authority to establish policies and strategies. Currently, the power and authority of many CDOs is evolving from data policy toward business strategy.

The Three Dimensions of the CDO Role

To provide more structure and a better understanding of CDO roles, we identified three key dimensions, as shown in Figure 1: (1) Collaboration Direction, (2) Data Space and (3) Value Impact.

1. Collaboration Direction Dimension: Inwards vs. Outwards

The Collaboration Direction dimension captures the focus of the CDO's engagement, either inside or outside of the organization. Collaborating inwards means focusing on internal business processes associated with internal business stakeholders. In contrast, collaborating outwards means that the CDO's focus is on stakeholders in the external value chain and environment, such as customers, partners, suppliers or regulatory bodies.

Initiatives led by internally focused CDOs typically include developing data-quality assessment methods or mechanisms; cataloguing data products, sources and standards; creating processes for managing metadata or master data; engaging in information-product mapping; and establishing data-governance structures. These initiatives seek to deliver consistent data inside the organization and to address the root causes of data-quality issues. Streamlining the internal business process associated with key data flows requires cross-functional cooperation, and can result in efficient and effective business operations. The CDO's success in these initiatives

---

5   Company examples and discussions on managing information as product vs. by-product can be found in Lee, Y. W., Pipino, L. L., Wang, R. Y and Funk, J. D. *Journey to Data Quality,* MIT Press, 2006.

6   Fernandez, T. "Chief Data Officers Becoming Crucial: GoldenSource," *Securities Technology Monitor,* June 18, 2012; available at http://www.securitiestechnologymonitor.com/news/GoldenSource-Chief-Data-Officers-30775-1.html.

7   See the Appendix for a brief description of the interviews conducted.

Figure 1: Three CDO Dimensions

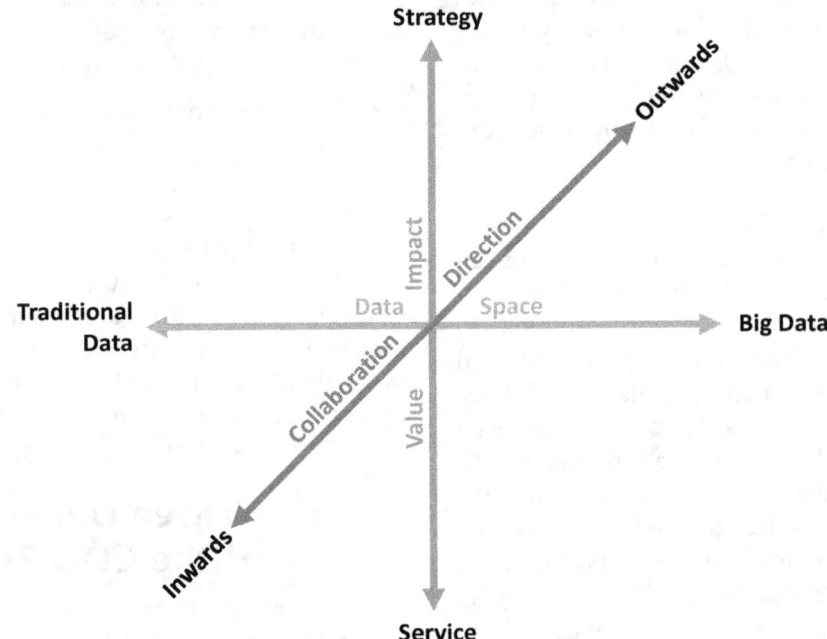

depends heavily on the ability to effectively lead the relevant internal stakeholders and map out the transformation journey.

In contrast, outwardly focused CDOs strive to persuade and collaborate with external partners. For example, the outwardly focused CDO of a global manufacturing company led a business-process-embedded "global unique product identification" initiative, aimed at improving collaboration with external global partners. Such CDOs may also focus on external report-submission activities, particularly if the company has experienced an external embarrassment or a sizable disaster created, for example, by poor-quality reports.

2. Data Space Dimension: Traditional Data vs. Big Data

The Data Space that a CDO focuses on can either be transactional data, typically in relational databases, or the newer and more diverse big data.

Many CDOs focus on traditional data, as it is the backbone of the organization's operations. Without a strong foundation in traditional data, an organization's most basic capabilities are hindered, and thus the need arises for a CDO focused on traditional data-management activities.

In contrast, big data is usually not connected with the organization's transactional data or database systems, but offers innovative opportunities to further improve operations or develop new business strategies based on new insights that traditional data cannot provide. CDOs focused on big data provide the leadership to adapt to and manage the analysis of this new, diverse type of data and to gain insights from these analyses.

3. Value Impact Dimension: Service vs. Strategy

The CDO's role can focus on improving services or on exploring new strategic opportunities for the organization. This dimension reflects the impact desired from a CDO. In many cases, the CDO role is a direct response to the on-going need for executive oversight and accountability to improve existing organizational functions. Increasingly, however, organizations require CDOs who can add strategic value by taking advantage of new tools such as data aggregators[8] or other data products based

---

8 Madnick, S. and Siegel, M. "Seizing the Opportunity: Exploiting Web Aggregators," *MIS Quarterly Executive* (1:1), 2002, pp. 35-46, explains web aggregators and their strategic business opportunities.

Figure 2: The Eight CDO Roles

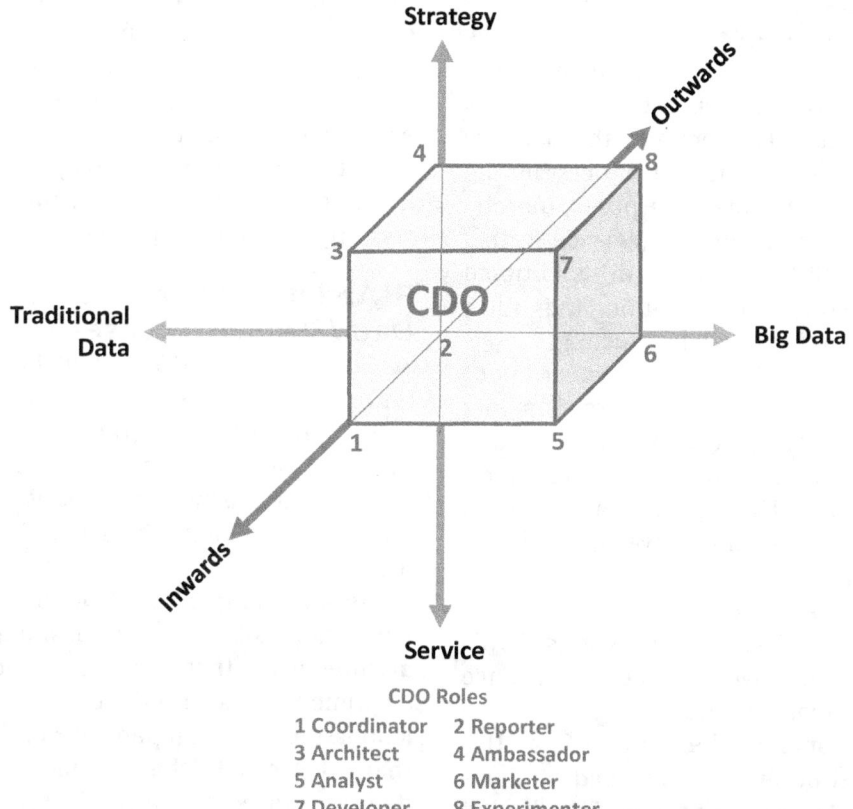

CDO Roles

1 Coordinator	2 Reporter
3 Architect	4 Ambassador
5 Analyst	6 Marketer
7 Developer	8 Experimenter

on digital data streaming.[9] These CDOs also explore ways to develop new market niches, or to transform the company so it can develop smarter products and services.

For example, one strategy-focused CDO led an initiative to identify new information products and advance the company's position in the financial industry. This CDO led a cross-organizational collaboration initiative to create a strategic vision for managing the new information products at the enterprise level. We have observed that CDOs who are positioned higher in the organization are better suited for taking on a strategy-focused role.

CDO Role Profiles

We have identified eight different CDO role profiles based on the three dimensions described above. These roles correspond to the eight corners of the CDO cube depicted in Figure 2.

For convenience, we have labeled the eight roles as "Coordinator," "Reporter," "Architect," "Ambassador," "Analyst," "Marketer," "Developer," and "Experimenter." "Coordinator," for example, corresponds with the corner defined by Inwards on the Collaboration Direction dimension, Traditional Data on the Data Space dimension and Service on the Value Impact dimension.[10] However, these names should not be taken too literally; they are simply a short-hand notation for each of the corners. Each of the eight roles is explained below.

It is important to note that, at any one time, a CDO may take on multiple roles. However, a CDO inevitably has one primary role. Moreover, it is common for a CDO to take on several different primary roles over time during his or her tenure as a CDO. Many CDOs that we interviewed noted that the evolution of their primary role was triggered by changes in the environment or the broader marketplace, as described below.

---

9 Picoli, G. and Pigni, F. "Harvesting External Data: The Potential of Digital Data Streams," *MIS Quarterly Executive* (12:1), 2013, pp. 53-64, explains new value-creating opportunities from digital data streams. One of the five value archetypes is aggregation of digital data.

---

10 Note that "Coordinator" is much shorter than saying "Inwards Collaboration direction dimension, Traditional Data Space dimension and Service Value Impact dimension."

1. Coordinator CDO: Inwards/ Traditional Data/Service Focuses

The Coordinator CDO manages enterprise data resources and sets up a framework that optimizes collaboration across internal business units (inwards focus). This enables the delivery of high-quality data to data consumers in the organization for their business purposes, thereby improving business performance (service focus). The Coordinator CDO works with traditional data, such as customer information and other transactional data (traditional data focus).

For example, the CDO at a U.S. government agency identified common critical data elements across the enterprise; these elements provided the foundation for data sharing and integration at the agency level. The agency then led an initiative to identify authoritative sources for these critical data elements. This work on common data elements set the stage for other data-improvement initiatives. Part of this CDO's responsibility was to oversee the governance process for data management.

In another example, the CDO of a U.S. healthcare institution established data-governance councils and workgroups. She also led the group responsible for enterprise-wide data quality assessment and improvement initiatives.

2. Reporter CDO: Outwards/Traditional Data/Service Focuses

In heavily regulated industries, such as finance and healthcare, an emerging trend in the CDO role is a focus on enterprise data to fulfill external reporting and compliance requirements. Like the Coordinator CDO, the Reporter CDO fulfills a business obligation (service focus) through the delivery of consistent transactional data (traditional data focus). However, the Reporter CDO's ultimate goal is to deliver high-quality enterprise data services for external reporting purposes (outwards focus).

For example, the CDO-equivalent at a U.S. healthcare institution oversaw the preparation of a selected set of data for regular reporting to the state government. She worked closely with other corporate officers, such as the chief medical officer and chief financial officer, as well as with external officials, to ensure that reports were delivered in a timely manner and that they accurately and effectively represented the activities of the institution.

Similarly, Reporter CDOs are often found in financial service organizations, working with compliance or risk-management groups to fulfill external reporting requirements. Typically, these CDOs are established when the company has experienced difficulties in producing external reports, and often they play an important role in integrating the data and information silos of recently merged companies.

3. Architect CDO: Inwards/Traditional Data/Strategy Focuses

The Architect CDO's Collaboration Direction and Data Space are the same as the Coordinator CDO (inwards and traditional data focuses), but the value impact comes from using data or internal business processes to develop new opportunities for the organization (strategy focus).

As an example, the CDO of a data company was responsible for establishing an enterprise architecture that would yield value-added customer data products. Under the CDO's leadership, the company developed a blueprint that described the business processes for delivering a new data product, the time required for each process and the individual responsible for each process. This blueprint, which we call the "map,"[11] was used to encourage members of the organization to collaborate on a daily basis. This CDO recalled: *"We made [the map] everybody's map. Everyone knows their data role in the company."* Suggestions for improvement to data products were also attached to the "map." This CDO reported that the "map" reduced time to market for new products by 50%. In addition, the company produced better data products, and did so before competitors could, thus gaining strategic advantage in the market.

4. Ambassador CDO: Outwards/ Traditional Data/Strategy Focuses

The Ambassador CDO promotes the development of inter-enterprise data policy for business strategy and external collaboration (outwards and strategy focuses) and focuses on traditional data. For example, the CDO in a financial services institution defined common datasets for risk management. He promoted a set of data standards and data-assessment measures

---

11   At the request of the company, we have used a pseudonym for the specific artifact.

for financial data exchange among peer financial institutions.

A second example cimes from an international bank in South America, which went through a strategic transformation that required significant process improvements and the establishment of data-governance mechanisms. During the transformation, the CDO, reporting to the CFO, led a close collaboration with other financial institutions to improve data security for electronic international money transfer processes and information exchange. This transformation was critical for the bank's business strategy and opened up opportunities to provide its customers with new services, which were previously not possible due to data-security weaknesses.

5. Analyst CDO: Inwards/Big Data/Service Focuses

The Analyst CDO resembles the Coordinator CDO, except that he or she focuses on improving internal business performance by exploiting big data, thus requiring different data-management and data-analysis capabilities. The need for an Analyst CDO often emerges after an organization hires data analysts or data scientists but does not assign an executive leader to provide an enterprise perspective for their efforts.

For example, a credit card company established a CDO who was responsible for overseeing internal teams evaluating and analyzing big data, such as geo-tagged data about credit card use and data from online customer surveys. This CDO collaborated with the chief risk management officer and provided direction for the data scientists. Subsequently, the company implemented enterprise-wide policies to improve risk management and fraud detection.

6. Marketer CDO: Outwards/Big Data/Service Focuses

The Marketer CDO develops relationships with external data partners and stakeholders to improve externally provided data services using big data. Marketer CDOs are often found in data product companies, where they develop working relationships with retailers, financial institutions, and transportation companies that are purchasing their companies' data.

For example, the CDO of a data product company worked closely with the company's customers, in this case healthcare institutions, to help extract insights from big data in the form of unstructured patient feedback data. This Marketer CDO led the analysis of the data to identify ways to alleviate key weaknesses of the healthcare institutions. While few CDOs may currently fulfill this role, we observe that the Marketer CDO is an emerging trend that is important for managing supply chain partners and customers.

7. Developer CDO: Inwards/Big Data/Strategy Focuses

The Developer CDO interfaces and negotiates with internal divisions to develop new opportunities for the organization to exploit big data. For example, the CDO in a retail organization worked with the chief marketing officer to find opportunities for new products and services based on mining consumer behavior data from geo-tagging along with consumer feedback data taken from social media sites. Using this vast source of information, this Developer CDO developed a personalized marketing strategy for the company.

8. Experimenter CDO: Outwards/Big Data/Strategy Focuses

The Experimenter CDO engages with external collaborators, such as suppliers and industry peers, to explore new, unidentified markets and products based on insights from big data. Through strong collaborative relationships across industries, this type of CDO maintains access to various sources of data and uses them for creating new markets and identifying innovative strategies for organizational growth.

For example, the CDO of a financial institution experimented with developing marketable information products for the broader financial industry and its prospective clients. In preparation, this Experimenter CDO suggested creating new information products by transforming, integrating and reusing data from multiple sources of consumer-generated data. More importantly, he presented this new product concept to the organization's clients to gain their feedback. This Experimenter CDO subsequently developed information products based on various data sources and marketed them to client organizations. He argued: *"We should be a revenue center, not a cost center."* By taking advantage of insights from the organization's diverse datasets and guided by his knowledge of shared industry needs, he expanded the organization's

Figure 3: Example CDO Role Evolution

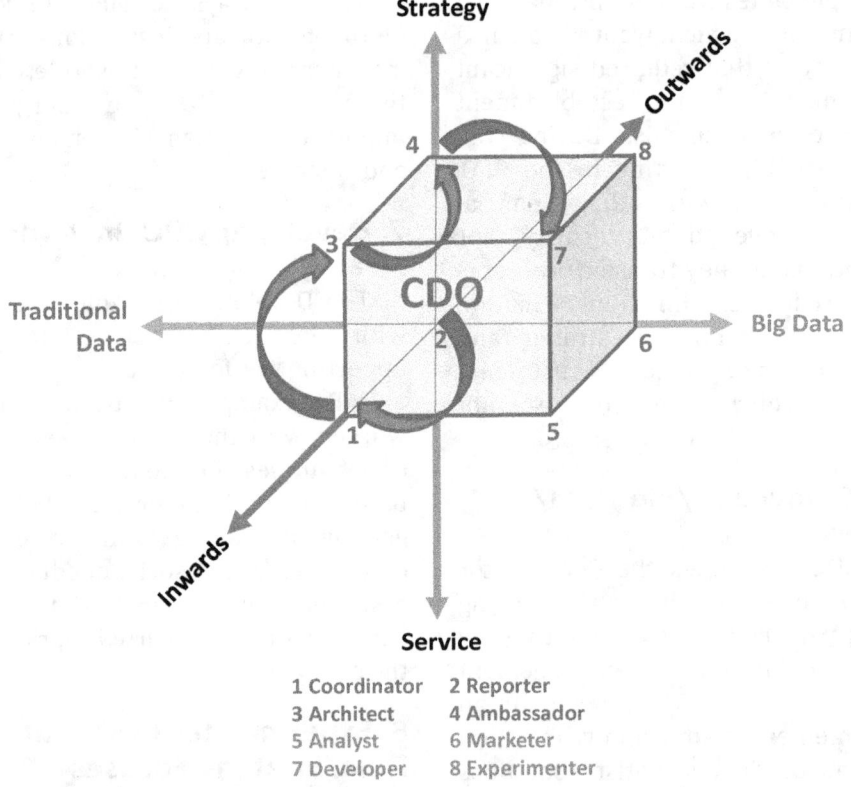

1 Coordinator 2 Reporter
3 Architect   4 Ambassador
5 Analyst   6 Marketer
7 Developer   8 Experimenter

capability to conceive and experiment with new information products, thus adding strategic value.

Example of the Evolution of the CDO Role

Not all businesses have the same needs and priorities, and thus the role of the CDO differs from company to company. Moreover, the role of the CDO can change as the needs of the organization change.

Figure 3 depicts how the role of the CDO at a U.S. hospital evolved over a period of 10 years. In this case, the CDO started with a focus on providing good service to external recipients of traditional data. Gradually her role took on a more strategic focus, both internally and externally, and presently she is concerned with exploiting big data. Over the 10 years we studied this institution, the CDO's role evolved from Reporter (Role 2), to Coordinator (Role 1), to Architect (Role 3), to Ambassador (Role 4) and now to Developer (role 7). Below, we briefly discuss this CDO's role over time and explain:

1. What triggered or prompted the CDO to transition to a new role

2. Why that role was chosen

3. What was accomplished by carrying out the new role.

1. Reporter CDO Role

Initially, the CDO fulfilled the Reporter role. As such, she oversaw the provision of data to state regulators, especially for reimbursements, since these were essential to the business. This was a challenge because the data, generated internally from the hospital's operations, often was not suitable for external reporting purposes. There were multiple sources of the same or similar data, producing inconsistent results. Several data sources were not trusted by internal data consumers, and thus some groups in the organization were reluctant to release that data for external purposes without further review. Every time there was a need for external reporting, the CDO had to go through all of the data, cleaning it up and preparing it for external submission.

2. Coordinator CDO Role

After being fined for submitting poor-quality data to the state government, the hospital realized that, to report good-quality data externally, it needed to turn its attention to internal data quality. Given a mandate from the CEO to improve the quality of organizational data, the CDO transitioned from the Reporter role to fulfilling the Coordinator role. She established an enterprise-wide data-quality improvement framework, coordinating across functional business units to systematically address *"cleaning up and preparing the data for submission."* In addition, she developed procedures to assess data-quality techniques periodically and established enterprise-wide dashboards for identifying and resolving data problems. Internal data consumers subsequently felt they could trust their data sources, and the external reporting process was also streamlined.

3. Architect CDO Role

Having successfully improved organizational data both for internal and external services, the CDO realized that there should be a sustainable structure and capability for data practices. This realization prompted her to fill the gap in sustainability by strengthening the alignment of data practices with business processes, thus changing her focus from service to strategy and assuming the Architect CDO role. In this role, she established governance for data quality, as well as standards committees and working groups. She also established and maintained an enterprise level data quality problem-solving process and aligned business roles with data roles for all members of the organization. She implemented a policy of assigning a specific data role to each member of the organization, such as a data collector, data custodian or a data consumer, in addition to a business role, thus strengthening business-data alignment. To reinforce the importance of data roles, each member's contribution to the quality of enterprise data was factored into their annual bonus.

4. Ambassador CDO Role

Increased pressure from insurance companies for comparable measurements required the CDO to improve collaboration between institutions. The CDO thus evolved to the Ambassador role in which she engaged in industry benchmarking and established shared data practices through a consortium and various forums. She participated in setting the industry's data roadmap, organizing and training other data practitioners and collaborating with other institutions to promote data quality across other hospitals. Through these efforts, the CDO transformed standards-setting for business processes and for various healthcare industry indices.

5. Developer CDO Role

The hospital's performance from the use of its internal data eventually reached a plateau. As a consequence, the CDO took on the Developer role, where she explored the use of big data generated by patients to improve hospital performance. In particular, she focused on developing various methods for analyzing unstructured patient feedback data to identify specific factors associated with poor performance. These analyses included data-mining techniques such as sentiment analysis. In combination with analyses of standard numerical assessments, such as the Hospital Consumer Assessment of Healthcare Providers and Systems report, the methods that the CDO developed led to actionable recommendations for doctors, nurses and other units within the hospital. In further pursuing such opportunities, the CDO is currently developing new measurements to provide more tailored feedback to the clinical teams for improving patient care and safety.

Guidelines for Using the Cubic Framework[12]

Our cubic framework can be used to identify the focuses an organization's CDO should have and hence the CDO role profile, a key to successful data practice. Below we provide a pragmatic three-step guide, based on the framework. In summary, the three steps are:

---

12 The authors benefited greatly from the advice, discussion and input from the *MIS Quarterly Executive* workshop on December 15, 2012, in Orlando, Florida.

- *Assess* the current status of your organization's data-related business practices (based on the three dimensions of the CDO cube)

- *Determine* the CDO role profile needed for your organization (based on the eight roles described), and whether an executive-level CDO is required to fulfill these needs

- *Strategize* the likely path for the CDO based on a projection of organizational future needs.

Step 1: Assess the Current Status of Your Organization

Assessing the current status of your organization's data-related practices will help to highlight the weaknesses you should focus on. The CDO cube provides a framework for identifying an organization's current needs with respect to the Collaboration Direction (inwards vs. outwards), Data Space (traditional data vs. big data) and Value Impact (service vs. strategy) dimensions.

In Table 1, we provide 12 assessment statements based on the cubic framework. Each statement is assessed on a seven-point scale (ranging from 1 [strongly disagree] to 7 [strongly agree]). Statements 1-4 relate to the Collaboration Direction dimension; statements 5-8 address the Data Space dimension; and statements 9-12 investigate the Value Impact dimension. To illustrate the assessment process, we have also included sample responses in the two rightmost columns.

Note that most organizations have needs that apply to every corner of the CDO cube; the responses to these assessment statements will help prioritize which roles (i.e., corners of the cube) are most critical. Responding to the statements is also an excellent opportunity to engage many members of the organization at all levels from a variety of business units. The varied perspectives will inform discussions about what CDO role is needed and will ensure the CDO has organization-wide endorsement.

Table 1 can be used both quantitatively and qualitatively. A simple quantitative analysis involves assigning a score (on a seven-point scale) for each response. Comparing the sum of the first two scores and the last two scores

for each dimension will reveal a bias in each dimensional space. In our example, statements 1 and 2 (emphasizing inwards) each have scores of three, and statements 3 and 4 (emphasizing outwards) have scores of six and seven. The sum of statements 1 and 2 (6) is less than the sum of statements 3 and 4 (13), suggesting that collaborating inwards is less critical than collaborating outwards. This same process can be repeated for statements 5-8 to determine whether the focus should be on traditional data or big data, and for statements 9-12 to determine whether the focus should be on service or strategy. Taken together, these comparisons give an indication of which CDO role is the most critical.

A qualitative analysis considers the "why" in the "Assessment Discussion Notes" column for each of the statements. This helps to determine the criticality of each dimensional direction. The examples shown in the rightmost columns of Table 1 are very terse; more comprehensive notes could be used for further elaboration.

Step 2: Determine Whether a CDO is Needed

Based on the assessment of its current status, an organization can move on to Step 2, which determines the CDO role profile needed and whether an executive-level CDO is required to fulfill those needs. Note that considerable discussion may be required before an organization can decide which roles are most important; the scores from Step 1 should not be taken as an immediate solution. Rather, the responses to the assessment statements should be used as a tool to initiate conversations among members of the organization on data practice and the implications for the organization's overall performance.

Establishing a new CDO role requires serious consideration because it implies a change in resource allocation and reporting relationships. Before establishing a CDO position, an organization should therefore review the effectiveness of other data-practice mechanisms, such as governance committees, workgroups and mechanisms for resolving data and business process conflicts. On the other hand, data-practice initiatives alone, without assigned accountability, often do not yield effective results.

Table 1: Example Assessment of CDO Role Based on the Cubic Framework

	Assessment Score (1-7) 1 Strongly disagree 4 Neutral 7 Strongly agree	Assessment Discussion Notes Why? Explain reason for assessment score
Collaboration Direction Dimension: Inwards vs. Outwards *High score for Nos. 1 and 2 implies inwards direction.* *High score for Nos. 3 and 4 implies outwards direction.*		
1. It is critical that our organization improves the effectiveness of data use for internal business operations.	*3*	*We do this well, thus not critical at this point.*
2. Our company has the opportunity to significantly improve internal operations.	*3*	*Maintain what we do well.*
3. It is critical that our organization collaborates with other value chain enterprises, such as suppliers, customers, distributors or competitors.	*6*	*We need to know our suppliers and customers much better.*
4. Our organization's success is critically interlocked with other companies, market changes, external situations or environments.	*7*	*Our procurement can be vastly improved with better understanding of our suppliers.*
Data Space Dimension: Traditional Data vs. Big Data *High score for Nos. 5 and 6 implies traditional data; high score for Nos. 7 and 8 implies big data.*		
5. Our organization's transactional data should be more effectively used to address the enterprise's needs.	*6*	*We need to know more about aggregated amounts of materials for different suppliers.*
6. It is critical for our organization to use transactional data in an integrated fashion across different business areas.	*7*	*To negotiate with our suppliers, we must get all divisions to use our existing information in a consistent way.*
7. Our company needs to identify opportunities for using big data and data analytics.	*5*	*We may not be there yet to go for this direction.*
8. It is critical for our organization to understand external sources of data, such as social media, for engaging customers.	*6*	*Our customers might be ready for new sources in the future, and we need to explore and exploit social media.*
Value Impact Dimension: Service vs. Strategy *High score for Nos. 9 and 10 implies service focus; high score for Nos. 11 and 12 implies strategy focus.*		
9. Our organization's data efforts should be focusing on maintaining the current needs of the business units.	*4*	*We are doing okay in serving the business units.*
10. It is critical for our organization's operations that we improve the efficiency of the data service.	*5*	*We can still improve, but we do well on serving data for the internal business units.*
11. Our organization's data efforts should be largely initiated by the need for changes in the way we do business.	*6*	*We can use the data for changing the way we do procurement planning with our global suppliers.*
12. Our organization must achieve its strategic business goals with better data.	*7*	*We must figure out who our best business customers are and set different strategies for different customers.*

Additionally, in some cases, organizations may already have leaders who can take on the role, or parts of the role, of a CDO. For example, the CFO may be able to take on the responsibilities that the assessment carried out in Step 1 would assign to a Reporter or Coordinator CDO, in which case a focus on traditional data and service may not be as critical as the assessment might suggest. We have also seen a case where a chief marketing officer has taken on the responsibilities of a Developer or Experimenter CDO role. In this organization, there was effective collaboration among senior executives, and in such cases, establishing a separate CDO role may not be necessary. More often, however, data-related collaboration among executives can be short-lived and ad hoc, and there is a need for the sustainable leadership made possible by a CDO.

Step 3: Strategize the CDO Evolution Path

Strategizing for future needs can be broken down into two processes. First, the organization should create a projected timeline for addressing the needs identified in Steps 1 and 2. For example, as illustrated in the rightmost columns of Table 1, an organization might determine that the primary need is for an Ambassador CDO role (outwards, traditional and strategy focuses). In this situation, the organization may create an 18-month plan to closely align data practices with business processes.

Second, based on quantitative and qualitative measures, the organization can determine how crucial other CDO roles in the cubic framework are relative to the primary role. Alternatively, the organization may determine that there are no other highly critical needs that must be addressed at this time. In either case, based on the projected timeline, the organization can either determine that the planned CDO will need to transition from one role to another, or it can decide to reassess organizational needs by repeating Steps 1 and 2 in the future.

In the example in Table 1, the scores for statements 5-8 suggest a small bias toward traditional data rather than big data (13 vs. 11). However, further analysis might suggest that big data demands are almost as critical as the traditional data needs that the future Ambassador CDO will be addressing. The organization could therefore plan for the CDO to evolve from the Ambassador role to

Experimenter role (outwards, big data and strategy focuses) at the end of the 18 months to address external needs.

An implicit, yet key strength of the three-step process is that it is a collective endeavor that engages all business units and functions. Enterprise support and approval for the establishment of a CDO lays the groundwork for the CDO to be an effective leader.

Concluding Comments

As organizations' strategies for achieving success increasingly depend on data, they must position themselves to harness the value of data. To this end, a growing number of businesses and government institutions are establishing CDO positions to exploit the critical value that data can provide. The three dimensions of the CDO cube framework presented in this article provide a guide for organizations as they analyze the need for a CDO and will enable them to determine the most appropriate profile for their CDOs now and in the future.

Appendix: Research Methodology

The study was conducted using three research methods: (1) initial informal case studies with multiple organizations; (2) detailed iterative interviews; and (3) structured surveys.

First, we used longitudinal informal case studies with 12 different organizations spanning various industries, including healthcare, finance, government, insurance, manufacturing, retail and IT. As part of our ongoing research on data practices, between 2003 and 2013, we conducted face-to-face interviews and on-site observations of these 12 organizations. The data we collected provided background on emerging CDO practices in the context of these organization, as well as their industries and the broader environment.

Second, during 2010-2013, we focused specifically on the CDOs of these 12 organizations. This entailed iterative interviews and semi-structured surveys, both on- and off-site, as well as continued onsite observation. For a comprehensive understanding of the CDO's work in the context of the organization, we also interviewed other executives and managers working directly with the CDOs on data quality, governance, data architecture and data strategies. The interviews were semi-structured and open-ended, typically lasting one and a half hours. In total, we interviewed 65 individuals—12 CDOs, 25 other executives and 28 managers.

Third, we developed structured surveys to collect concrete and more detailed statistics on organizational practices relating to CDOs, such as reporting relationships. Between 2010 and 2013, we surveyed 95 CDOs and data practitioners and collected a wealth of data from which we could tease apart different patterns and rules of CDO practice.

Together, these three methods provided a detailed and comprehensive picture of the contemporary data practices of the chief data officer. The longitudinal research provided critical context for the study; the focused interviews with CDOs provided the activity-level details needed for devising the cubic CDO framework; and the surveys provided the statistical power to identify key trends of the CDO role.

About the Authors

Yang W. Lee

Yang Lee (y.lee@neu.edu) is Associate Professor at Northeastern University, D'Amore-McKim School of Business. She was founding Co-Editor-in-Chief of the *ACM Journal of Data and Information Quality* and co-founder of the *International Conference on Information Quality*. Her research focuses on information quality, problem solving and institutional learning, and strategic use of information. Her publications have been widely cited, translated into various languages and applied globally in the public and private sectors. She has received numerous recognitions, including the 2012 DAMA International Achievement Award and the 2005 Certificate of Appreciation from the Director of Central Intelligence, U.S. She received her Ph.D. from MIT.

Stuart E. Madnick

Stuart Madnick (smadnick@mit.edu) is the John Norris Maguire Professor of Information Technology in the MIT Sloan School of Management and Professor of Engineering Systems in the MIT School of Engineering. He received his M.B.A. and Ph.D. in Computer Science from MIT, has been an MIT faculty member since 1972 and has headed the IT group for more than 20 years. He is Co-Director of the PROductivity From Information Technology and Total Data Quality Management research programs. He is the author/co-author of over 300 books, articles or reports. His research interests include integrating information systems, data quality and strategic use of IT.

Richard Y. Wang

Richard Wang (rwang@mit.edu) is Director of the MIT Information Quality and the MIT Chief Data Officer Research programs. He was a professor at MIT Sloan School of Management for almost a decade. He has also served as the Deputy Chief Data Officer and Chief Data Quality Officer of the U.S. Army. He is the recipient of numerous awards, including the DAMA International Achievement Award, the German Society of Information Quality Award and the International Association of Information and Data Quality Award. Wang received his Ph.D. from the MIT Sloan School of Management.

Forea L. Wang

Forea Wang (forea@stanford.edu) is a Ph.D. candidate at Stanford University School of Medicine in the Neurosciences Program. She is a recipient of the Stanford Graduate Fellowship, the National Science Foundation Graduate Research Fellowship and the MIT Information Quality Program's Decade of Outstanding Contribution Award. She received her B.S. in biological engineering from MIT.

Hongyun Zhang

Hongyun Zhang (zhanghongyun@mail.xjtu.edu.cn) is a post-doctoral fellow at the Center of Data Science and Information Quality, School of Management, Xi'an Jiaotong University, China. In addition, she is also a visiting research scholar at supply chain and information management group, D'Amore-McKim School of Business, Northeastern University. From 2008 to 2009, she was a visiting scholar in Manchester Institute of Innovation Research, Manchester Business School, University of Manchester, U.K. Her research interests include the chief data officer role, data quality and entrepreneurial orientation.